THE EARTH BENEATH US

THE EARTH

A Chanticleer Press Edition

Photographs by

Josef Muench
Mary and Benjamin Shaub
Robert Clemenz
Andreas Feininger
Emil Schulthess
Steve McCutcheon
Bradford Washburn
Tad Nichols
Gerhard Klammet
and others

Drawings by Howard Morris

BENEATH US

The Random House Illustrated Science Library

by Kirtley F. Mather

Professor of Geology, Emeritus,
Harvard University

Random House, New York

Planned and produced by Chanticleer Press, New York

*Special map credits: pages 98 and 134, adapted from "The
Physiographic Provinces of North America" by Wallace
W. Atwood, 1940; pages 182–183, after J. S. Hough,
in "The American Scientist," March 1963; pages 184–185,
adapted from "Glacial and Pleistocene Geology"
by R. F. Flint, 1957.*

Library of Congress Catalog Number: 64–17101

Manufactured in the United States of America

Contents

Preface

Recalling how Antæus in his stupendous battle with Hercules drew his strength from the earth that was his mother, the goddess Gæa, it would be wise for us to keep in mind our complete physical and biological dependence upon earth.

Our dependence on the earth is not only physical and biological but esthetic and spiritual as well. The bonds that tie man to earth involve not only his food and drink, the air he breathes, and the materials for his shelter and tools, but also his estimate of his place in the natural scheme of things. It is easy for twentieth-century man to forget this relationship. Insulated from the ground by asphalt or concrete, restricted in outlook by the steel and glass canyons of his cities, supplied with goods made by chemists, he is getting farther and farther away from his natural environment. It is time that all of us became better acquainted with the earth on which we live and from which we live—in short, become, as much as we can, geologists.

This is a book for the "intelligent layman"—the person who is interested in his natural surroundings and eager to understand the world in which he lives, but who has not had any special training in the earth sciences. Even at the risk of oversimplification I have therefore tried to set forth the concepts of geology without using any more professional jargon than was necessary. This proved most difficult in Chapter Three, on rocks, because it deals with more technical matters than any other. The reader may perhaps want to scan this chapter and then return to it when he later discovers how important different kinds of rocks are in determining the features of many landscapes.

The pictures are just as important as the text. Each photograph has been selected because it has something to tell about the way the world was made. Most of them show phenomena that are repeated many times on every continent; the localities pictured were therefore chosen from far and wide. If the reader will compare with them the natural features he encounters wherever he may travel he will be well on his way toward seeing the world as geologists see it.

I am deeply grateful to Chanticleer Press for making it possible for me to fulfill a long-nurtured ambition to produce a "pictorial geology for the layman" and especially to the editors, Mr. Milton Rugoff, Mrs. Patricia Lasché Zunz, and Mr. Ulrich Ruchti for their patience with me. Mr. Rugoff was especially skillful in getting me to simplify my language and helping me to tighten up the text. Mrs. Zunz devoted untold hours to the arduous task of securing and screening the many thousand photographs from which the few hundred used here were chosen. To Mr. Ruchti goes the credit for the attractive layout and design.

My hearty thanks are also extended to my former students and colleagues who contributed photographs that could have been taken only by individuals who combined the knowledge of the expert geologist with the technical skill of the expert photographer. Their names appear in connection with the pictures they supplied. But most of all, I want to express my gratitude to my wife, Marie Porter Mather, for continuing encouragement, wise advice concerning semantic problems, and unstinted zeal in typing and retyping the manuscript.

Kirtley F. Mather

Cambridge, Massachusetts
January, 1964

Preface to the Second Edition

Since 1964, the science of geology has grown apace. New tools and techniques have been made available, chiefly by oceanographers, seismologists, and geophysicists. Radically new concepts, such as plate tectonics and the spreading of ocean floors, have stimulated lively discussions and investigations. An unprecedented number of men and women have enlisted in disciplines within the earth sciences and their output of research papers has become almost overwhelming. Financial support has come in record-breaking amounts from governments and institutions around the world. It can therefore truly be said that geology is alive and well in this last third of the twentieth century.

Just as important is the increased interest in earth sciences as a result of deepening concern for human welfare on "spaceship earth" and the realization that there can be no satisfactory ecological adjustment for mankind or conservation of resources without an understanding of geological processes and earth history. I therefore welcomed the opportunity to bring this text up to date.

Albuquerque, New Mexico   *Kirtley F. Mather*
January, 1975

I

Getting Acquainted with the Earth

Most people think of the earth beneath us as something that never changes. Indeed, ever since the days of the ancients, who spoke of it as *terra firma,* the earth has been a symbol of firmness and eternal stability. But to a modern geologist the earth is neither stable nor inert but alive, its exterior changing constantly, sometimes with explosive or cataclysmic violence.

To get some idea of a geologist's view of the earth and its history, an ideal instrument would be a time-lapse camera such as those used to photograph the progress of a flower from bud to full bloom. One can imagine a motion picture of the earth, taken by a camera on a platform some thousands of miles out from the earth's surface, with one picture of the same hemisphere taken every five thousand years. After nearly a billion years, we would have a movie of truly epic proportions, telescoping a large part of the history of our planet into a three-and-a-half-hour drama.

In such a motion picture the earth would appear to be alive, its exterior writhing in spasms. Great wrinkles—mountain ranges and canyons—would appear in land that a few moments before had been flat. Lands previously covered by shallow water would emerge and other lands would be flooded by spreading seas. Glaciers and running water would file the mountain ranges into jagged peaks, then down to low hills and finally back to flat valleys once more. Green jungles would change suddenly into stark deserts. Great gashes, such as the rift valleys in Africa, might open up in a few seconds. Volcanoes would fairly jump out of the surface and then be worn away in a minute or two. Vast ice sheets would expand over immense areas, then retreat and expand and retreat again, carving the land and leaving behind new rivers, lakes and soils.

This is the kind of rapidly changing world the geologists see. They do so by analyzing the short-term processes that can be observed even during an ordinary lifetime. Combining their observations, they then formulate their view of how the surface of the earth came to be as it is.

This geological knowledge bears fruit in many ways. For one thing, it yields rich material returns. The old-time prospectors used to say, "Gold is where you find it," but gold can be found only in rocks and minerals of particular kinds in certain geological localities. This is also true of the many other valuable metals such as silver, lead, copper, zinc, iron and aluminum, of the mineral fuels, such as coal, oil and gas, and now, in this age of atomic energy, uranium and thorium. A knowledge of the geological conditions associated with these substances is increasingly essential to the successful search for them. Throughout large parts of Europe and North America, the more accessible ore bodies and oil pools have already been found. But many are exhausted. The more completely concealed or more deeply buried stores will be found hereafter mainly by industrious, adventurous and, above all, adequately trained men. Most of the earth's resources are, in fact, available only to those who understand the ways of nature.

This of course implies the training and experience of a professional geologist. But even a modest acquaintance with geology helps one understand familiar landscapes or formations of earth and rock. The pleasure felt by the scientist when his research is successful can be shared by anyone who learns enough to be able to detect causes and effects. A landscape is far more interesting to eyes that are sharpened and senses that are alerted by understanding.

Beyond these utilitarian and intellectual fruits is the spiritual enrichment that comes from "conversation with the earth." When Job said, "Speak to the earth and it shall teach thee," he meant that it would teach far more than how to find metallic ores, coal seams and oil pools, or how to appreciate the beauty and grandeur of the world. Among its most valuable lessons—for this writer, at least—are those that strengthen the sense of oneness of man and nature, increase awareness of the marching orders of the universe and renew faith in the dignity of the human spirit. Here, for once, it is truly appropriate to say that there are "books in the running brooks, sermons in stones." Perhaps the distinction made in 1473 by the Bishop of Durham between "geologians" as those who deal with "worldly things" and the theologians as those who deal with "heavenly things" is not so absolute as the bishop thought it was.

THE GEOLOGIST'S VIEW

Of course, many of the early ideas about the composition and history of the earth were strongly tinged by theological views. Late in the fifteenth century, Leonardo da Vinci had great difficulty in persuading scholars that Noah's Flood could not have been responsible for the fossils of marine animals found in the layers of rock high in the mountains of Lombardy. And Athanasius Kircher (1602-1680) let his imagination run wild when he drew a fantastic cross section of the earth to portray "the compartments of heat or of fire, or what is the same thing, the fire cells, throughout all the bowels of the Geocosm, the wonderful handiwork of God."

The statement of basic principles for acquiring truly scientific knowledge about the earth dates back, however, to 1695. In that year, John Woodward, professor of "physick" in Gresham College, London, published "An Essay towards a Natural History of the Earth." The description of his procedure was a milestone along the path of the slowly advancing science:

> "Wheresoever I had notice of any considerable natural spelunca or grotto: any sinking of wells: or digging for earths, clays, marle, sand, gravel, cole, stone, marble, ores of metals, or the like, I forthwith had recourse thereunto: where taking a just account of

10

every observable circomstance of the earth, stone, metall, or other
matter, from the surface quite down to the bottom of the pit, I
enter'd it carefully into a journal, which I carry'd along with me
for that purpose. And so passing on from place to place, I noted
whatever I found memorable in each particular pit, quarry, or
mine: and 'tis out of these notes that my observations are
compil'd."

This is still a sound description of the scientific approach, especially because
Woodward's "observations" included not only descriptions of what he had
seen but explanations in terms of cause and effect. The geologically oriented
observer always asks, "How did what I see get to be the way it is?" The
scientist follows precise inspection with creative thinking that leads through
inference to the building up of general theories.

One of the most influential of such theories was set forth in 1795 by James
Hutton, a "private gentleman" of Edinburgh. Hutton's "theory of the earth"
is quite simple, but it provides guidelines for the imaginative thinking that
makes geology so successful as a servant of mankind. "No extraordinary
events (are) to be alleged to explain a common experience . . . Chaos and
confusion are not to be introduced into the order of nature, because certain
things appear to our partial views as being in some disorder. Nor are we to
proceed in feigning causes, when those seem insufficient which occur in our
experience." The geologic processes now modifying the earth have been
operating uniformly throughout its entire history, he asserted. Therefore he
urged the study of those processes as they may be observed today. Even
though changes during the lifetime of any one observer may seem slight,
when they continue throughout long periods they are sufficient to explain
all that we see.

The proof of Hutton's principle is not to be found in its logic, but whether
it works in practice; this is a test that is being applied a thousand times a day
in every well drilled for oil or gas or water, in every mine shaft sunk in
search of a valuable ore, in every groin built along a shore to protect it from
further erosion.

EXTERNAL FORCES THAT SHAPE THE EARTH

Some of the processes that change the earth's surface and have produced its
rocks and mineral resources are due to forces of nature operating on or near
the earth's surface. Other forces originate deeper in its interior.

The external agents of geologic change operate largely because the solid
earth is surrounded by a shell of gas, the *atmosphere,* and a discontinuous
shell of water, the *hydrosphere*. Both air and water are in almost constant
motion. They are heated by the sun and respond to the gravitational pull
of the rotating earth, setting the stage for the drama in which winds, waves,
running water and creeping glaciers play leading roles.

Anyone who looks closely at sand dunes with their wind-rippled slopes
becomes immediately aware of one of the activities of the atmosphere as a
geologic agent: the transportation of earth materials from place to place.
Although at first the wind may seem to blow "where it listeth," the rules

determining the location and forms of deposits of wind-blown sand are now well known. So are the processes whereby wind-driven sand erodes the rocks and often sculptures bizarre features in dry regions. Less conspicuous but more widespread is the chemical activity of the atmosphere; its moisture, oxygen, and carbon dioxide are always at work, even in calm air. This, combined with changes of temperature conditioned by the atmosphere, results in rock weathering, often the first step in the process of erosion.

The greatest contribution of the atmosphere to the continuing changes in the face of the earth is the part it plays in the "water cycle." Moisture evaporates from oceans, lakes and rivers, from damp ground and from the leaves of plants. The water vapor rises invisibly in air currents, condenses in clouds, and some of the clouds drift over the land. There the water falls as rain, sleet or snow, and runs from higher to lower ground and eventually to the sea. It is an endless cycle, ever renewing the supply of running water to erode the land.

Work done by running water is primarily responsible for the configuration of landscapes over much of the globe. The energy of a torrential stream on a steep slope is apparent to anyone. So also is the power of any river in flood. But it is a truly mind-stretching experience to contemplate the result of long-continued stream erosion in such a stupendous chasm as the Grand Canyon of the Colorado River.

Much of the rain falling on the land as well as of the water from melting snow is absorbed into the thirsty ground, soaking downward through the crevices and pores of the underlying rocks. In contrast to the surface runoff, this is known as "ground water." Because it supplies springs and wells, it is of prime importance in the life of man. As a geologic agent it also has great significance. Ground water ordinarily percolates slowly through minute pores and tiny passageways. Therefore it cannot move material by sheer force; its work is accomplished largely through chemical action. Almost every one of the famous caverns throughout the world is the result of the solution of limestone by ground water. In most of them the ground water has engaged in deposition as well as solution, forming deposits of great variety and beauty.

Ground water also forms hot springs and geysers, produces valuable metallic ores, and is responsible for certain features of the landscape in particular localities. Altogether it is one of the important agents of geologic change, even though its major work is done under cover and largely out of sight.

Eventually, much ground water returns to the sea by way of surface streams or through submarine springs. There it contributes its annual quota of dissolved mineral matter to the vast reservoir of soluble salts in the oceans. Although not apparent to the casual observer, these are becoming increasingly important to human welfare.

Almost everyone is aware of the role played by the waves and currents of oceans and lakes in changing the face of the earth. Waves frequently demonstrate their power along every shoreline and the explanation of shore features is thus not difficult. Knowledge of shore processes is often of great economic significance; land must be protected from erosion and harbors must be maintained or enlarged at many places.

Some of the hieroglyphs in the diary of the earth require a knowledge of the activity of snow and ice, which was much more widespread in the geologic past than it is today. Large areas now enjoying a temperate climate have soils and landforms that are a legacy of the Ice Age. The processes responsible can be observed in operation among glaciers and ice caps now in existence.

The last of the geologic agents operating essentially on the earth's surface is gravity. Landslides and rock avalanches have often destroyed roads, railways and buildings and have occasionally snuffed out many lives in great catastrophes. Other manifestations of the direct effect of gravity range from the well-nigh imperceptible downslope creep of soil on hillsides to the crashing collapse of a cliff undercut by wave erosion.

THE INTERNAL FORCES

Whereas there are numerous external agents of geologic change, the internal agents number only two: *vulcanism* and *diastrophism*.

Looking into the fiery pit of such a volcano as Kilauea in the Hawaiian Islands, one sees eddying lava that represents the top of a column of molten rock extending far down into the earth's crust. Deep below the surface, the rocks are changing under the influence of high temperature and hot gases. These subterranean effects are so important that the phenomenon is known as "vulcanism," a name derived from Vulcan, the Roman god of fire and metal-working, rather than "volcanism," which is limited to the eruptive activity of volcanoes. Volcanic eruptions are of interest not only as awesome natural forces, but also because they provide clues to what happened long ago when rocks now at the surface were deep in the earth's interior.

In a way, the displays of distorted rock layers seen in many mountainous regions are as spectacular as the eruption of a volcano. The process of distortion cannot be witnessed and we can only infer that forces within the earth's crust were strong enough to produce the extraordinary result. In this instance it is as though a segment of the crust had been squeezed horizontally in a gigantic vise. In other places, such forces have pushed plateaus upward or caused great blocks of the crust to sink downward or move laterally. This process is called "diastrophism"; it makes continents stand above ocean basins and the great mountain ranges appear as wrinkles on the earth's face.

Each detail of every landscape may be explained in terms of the operation of one or more of these various geologic processes. With them in mind, it is possible to reconstruct the sequence of events responsible for the landforms and the natural resources that have influenced so profoundly the activities of mankind throughout human history.

13

2

The Earth in Space and Time

In 1655 Archbishop James Usher of Armagh, Ireland, published a chronology of the earth based on computations from Biblical sources. The earth, he calculated, was created in the year 4004 B. C. in the manner described in the first chapter of Genesis. The archbishop believed, along with most people of his time, that the earth was the immovable center of the universe, although there were some who asserted that the earth revolves around the sun, and that the sun is at the center of the universe.

Although such ideas as Usher's dominated thought for many years after 1655, they were abandoned by scientists a long time ago. Knowing a great deal more about the earth and the universe than our ancestors, we also know a great deal less. That is, we have learned to be able to say that we simply do not know exactly when or how the earth came into being, but that it was billions of years ago and by a process different from that suggested in Genesis. Neither the origin of the earth nor its location in space is the direct concern of geology, but some understanding of both lends perspective to the subject. The age of the earth, on the other hand, is of basic importance to the geologist.

THE LOCATION OF THE EARTH

The position of the planet earth in space may be represented by the two pictures on pages 16 and 17. Each shows a galaxy photographed through the two hundred inch telescope on Mount Palomar; each galaxy is an aggregate of several billion stars and much interstellar gas and dust arranged in the pattern of a discus. From the earth, one of the galaxies in the constellation "Canes Venatici" is seen edge-on. The other is approximately at right angles to our line of sight. The "pinwheel" pattern of the latter indicates it is a spiral galaxy; presumably the other one is also.

The system of stars that includes our sun is known as the Milky Way galaxy. From a point in space, beyond the stars in the Milky Way, it would look almost precisely like the nebula on page 16; seen from the side, it would closely resemble the galaxy on page 17. On such pictures of the Milky Way, our sun would be a tiny, unobtrusive point of light about three-fifths of the way out from the center of the galaxy toward its edge. The earth, an infinitesimal speck of dark matter close by, would be utterly invisible even from the next star.

14

Unfortunately, there is no familiar object in either galaxy photograph to give any notion of their dimensions. Using the light year (the distance traversed in one year by light traveling at 186,000 miles per second) as a unit of space measurement, the distance of the earth from the center of our galaxy is approximately twenty-seven thousand light years. Translated into miles, this equals $27,000 \times 186,000 \times 60 \times 60 \times 24 \times 365$, giving a product obviously too great for our minds to grasp.

The longer diameter of the Milky Way galaxy is nearly four times as great, or about a hundred thousand light years; its shorter diameter is about twenty-five thousand light years. So large is the galaxy that it holds billions of stars, each several light years from its nearest neighbor. The star nearest the sun, for example, is four and a half light years away. Thus there is plenty of room for the earth and the other planets in the solar system to continue revolving around the sun without external interference.

Like the galaxy, the solar system occupies a discus-shaped region. If the orbits of the eight planets, from Mercury out to Neptune, are drawn to scale, concentrically around the sun, on a blackboard six feet square and a quarter-inch thick, each planet as it revolves would always keep within the quarter-inch space between the front and back faces of the board. The ninth planet, Pluto, with its orbit outside that of Neptune, is somewhat errant; the blackboard would have to be a couple inches thick to keep Pluto between its faces at all times. Such a thickness would also be necessary to prevent some of the asteroids—hundreds of tiny planets orbiting between Mars and Jupiter—from popping out occasionally in front of the blackboard or behind it.

In addition to occupying almost the same plane, the planetary bodies of the solar system all move counter-clockwise as viewed from the direction of the North Star. The sun also rotates counter-clockwise on its axis. Moreover, all the planets except Uranus rotate upon their own axes in the same direction. With few exceptions, all the satellites, including the Moon, also revolve around their controlling planets in that direction. These are among the many things that must be considered in attempting to explain how the earth was made.

ORIGIN OF THE EARTH

The earth was in existence long before the oldest rocks now observable were formed. Thus its composition and structure provide only scanty clues to its origin. The problem involves the beginning of the solar system as a whole rather than merely that of the earth. Unfortunately, no other system of planets is known. With a dozen other systems for comparison, a trustworthy theory of the origin of all could almost certainly be developed. But even the nearest star is too far away for us to detect any planets circling it.

Since there are many other stellar galaxies like the Milky Way, and many other stars, both in our own galaxy and in others, similar to the sun, there may be many millions of planetary systems scattered throughout the observable universe, some harboring intelligent life forms. This assumes that the formation of planets around stars like the sun is a normal occurrence, and there is no reason to believe that it is not.

Even so, our own solar system is the only one we know anything about.

15

A nebula in the constellation Canes Venatici, seen through a 200-inch telescope, appears very much as would our galaxy if viewed from far out in space beyond the stars in the Milky Way. (Mount Wilson and Mount Palomar Observatories)

In considering its origin, the best we can do today is to imagine turbulent, whirling clouds of gas and dust that develop eddies and sub-eddies of diffuse but swiftly moving matter. Under the influence of gravitational, magnetic and radiational forces, these may condense or coagulate to form stars like the sun, and to produce its family of planets and satellites. It remains an open question whether the earth condensed from a larger, essentially gaseous protoplanet, passing through a liquid stage to the present solid state, or whether the protoplanet was a smaller solid body that subsequently grew to its present dimensions by the accretion of solid particles. This latter view is attractive because the earth is still growing, albeit at a trivial rate; the fall of meteorites adds many tons to its mass each year. These may represent the final stages in a long task of "cleaning up" that portion of space swept by the earth in its annual circuit of the sun. Once more is known about the present structure and composition of the earth's interior, geologists may help to determine whether in its youthful state the earth was a liquid sphere, with approximately its present mass, or a much smaller solid body.

THE MOON AND THE TIDES

Contrary to the beliefs promulgated by some textbooks and popular science writing, the problem of the origin of the moon is still unresolved. The earth is unique in the solar system (and, so far as we know, in the universe) because of the relative dimensions of this planet and its solitary satellite— solitary, of course, only until men began putting artificial ones in orbit.

Titan, one of Saturn's nine known satellites, is larger than our moon; it has a diameter of about 3100 miles whereas the diameter of the moon is just 2160 miles. But the mass of Saturn is at least five thousand times that of its satellite, Titan, whereas the earth weighs only eighty-one times as much as its moon. Moreover, Titan is 760,000 miles from Saturn's center and the moon is on the average only 239,000 miles from the center of the earth. Consequently the mutual gravitational attraction between earth and moon produces tidal forces not duplicated anywhere else in the solar system.

One result of these tidal forces is the *equivalence* of the moon's periods of rotation and revolution: viewed from the earth, the moon does not appear to rotate at all, but permits us to see only one of its hemispheres. Another result is the slowing down of the earth's rotation through the long ages of its history. At present, our twenty-four-hour day is growing longer at the rate of fifteen ten-thousandths of a second per century. This may seem scarcely worth considering, but projected backward for a few billion years, the tidal retardation leads to a surprising conclusion. Long ago, the earth was rotating so swiftly that a day was only eight or ten hours long. At that time, the moon was less than fifty thousand miles away; since then, it has slowly moved outward to its present position—another result of earth-moon tidal action.

The above phenomena first seemed to suggest that earth and moon were originally one heavenly body and that the moon was "thrown out" or "pulled away" from one side of the earth. This led to the familiar supposition that the Pacific Ocean basin is the depression left by the moon's departure. This old idea is now generally rejected; modern knowledge of celestial mechanics makes it appear quite impossible, and leads inexorably to the conclusion that

A spiral nebula in Canes Venatici resembling our galaxy as it would appear if viewed from a direction at right angles to the plane of the Milky Way. (Mount Wilson and Mount Palomar Observatories)

17

the moon is as old as the earth, both having come into existence at the same time—a protoplanet and a protosatellite—in the early evolutionary development of the solar system. Whatever the earth's and moon's original dimensions may have been, they were probably close enough together so that, from the beginning, their mutual gravitation kept them in a more or less stable relationship as they revolved about their common center of gravity. If, as seems likely, the earth has grown larger by accretion of matter, then so too has the moon. Such growth would increase their mutual attraction, but the centrifugal effect of the moon's revolution around the earth has never failed to keep them apart. In fact, ever since the two bodies attained approximately their present mass, the tidal forces have caused the moon to recede more and more. This will continue until the earth's rotation is slowed down so much that the terrestrial day becomes exactly equal to the lunar month. But that will require many billions of years.

THE EARTH'S PLACE IN THE UNIVERSE

Primitive man quite naturally assumed that the earth was a stable platform from which he could watch the heavenly bodies move majestically over his head. The idea that man dwells on the surface of a rotating globe revolving around the sun was not easy to accept. Human senses do not casually perceive such movements. Only minds free to seek facts regardless of inherited traditions could become aware of the actual relations between the earth and the other bodies in space.

There is an oft-repeated story that when Galileo left the Chamber of the Inquisition in 1633, after repudiating the Copernican theory that the earth revolves around the sun, he whispered, *"Eppure si muove!"* (Even so, it moves!). The story is almost certainly false. But if it were true, Galileo was by no means prepared to accept the motions that are attributed to the earth today. He appreciated well enough the fact that he, like everything else in the latitude of Rome, was being carried along at better than sixteen miles a minute by the rotation of the earth on its axis and at the same time was rushing through space at more than one thousand miles a minute, as the earth moves in its orbit around the sun. But for him the sun was the immovable center of the universe. In contrast, it is now known that the sun and the whole solar system are moving in an elliptical orbit around the center of gravity of the Milky Way at a speed of at least eight thousand miles a minute, and the galaxy as a whole is moving through the inconceivably vast space of the universe at an even faster, though not precisely known, speed. If Galileo could review the data gathered by astronomers in recent years, he would be nearly as astonished by the scientifically perceived motions of the earth through space as are the rest of us, unaware as we are of what the earth is actually doing. Interestingly enough, many of the dwellers on the earth today have had experiences denied to Galileo and his inquisitors, which make it relatively easy to accept the idea of the earth's swift motion through space. Cruising in a jet plane at ten miles a minute does not really *feel* five times as fast as cruising in a piston-plane at two miles a minute. Our sense perception reports *change* of velocity, not velocity itself, unless there are nearby objects with which to relate our motion.

18

THE EARTH IN TIME

The recent revolution in the minds of men concerning the place of the earth in the universe has greatly influenced the study of astronomy. A similar intellectual revolution regarding the earth in time has had even more profound consequences for geology. Indeed it forms the basis for the study of the earth's history.

When Archbishop Usher computed that the earth was less than six thousand years old, his figures were widely accepted and were even inserted in the margins of many Bibles published in the nineteenth century. Such a brief time seemed to be adequate to encompass all known events in the history of the earth and of man. But times have changed.

Before the close of the eighteenth century, geologists became aware that many rocks must have originated millions of years ago. This discovery cleared up many puzzles for it "provided" a long time span within which geological processes occurred. In the nineteenth century, archeologists began to unearth records of prehistoric civilizations dating back thousands of years before written records and even older human artifacts. During that same century, astronomers joined geologists in attributing an age of at least a hundred million years to the earth and the solar system.

During the twentieth century, a method has been developed for measuring with precision the time consumed by the long history of the earth and the universe prior to the advent of man. The timekeepers are radioactive elements. Various forms of such elements decay at rates that do not vary with changes in pressure or temperature. There is every reason to presume that these decay rates have persisted throughout the life of these radioactive elements. Therefore the ratio between the amounts of decay products and radioactive parent materials provides the index to the time elapsed since the

The planets in the Solar System revolve around the sun in the same direction and with orbits approximately in the same plane. As shown here, Mercury, Jupiter and Mars are to the right of the sun, and Venus, Earth, Saturn, Uranus, Neptune and Pluto to the left (drawing not done to precise scale).

19

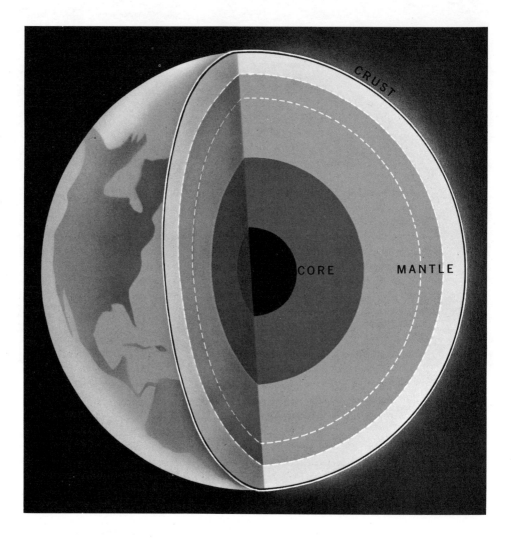

A cut-away of the earth showing (a) the inner core of solid iron and nickel with a radius of 800 miles, (b) the outer core of liquid iron and nickel extending an additional 1400 miles outward, and (c) the mantle of rocky material in three concentric zones extending to the crust. The crust is about five miles thick below sea floors but more than twenty miles thick in the continents.

particular substance came into being. Several sequences of radioactive decay have now been measured with precision. For example, uranium 238 changes in a series of stages to lead 206, thorium 232 eventually becomes lead 208, potassium 40 turns into argon 40, and carbon 14 to nitrogen 14. The antiquity of many events in the history of the earth may now be announced not simply relatively but in terms of years.

The oldest known rocks are in central Canada (the Canadian shield), western Australia, the Ukraine, and on the Kola Peninsula in the Soviet Union. Their ages range between two and a half and three and a half billion years. None represents an original crust of a solidifying earth. Rather, they are the roots of long-vanished volcanoes, lava flows that must have poured out on previously existing solid floors, or else ancient sediments collected in depressions on a solid surface. The earth must be appreciably older than the oldest of them.

To determine the age of our planet we must appeal to astronomers and especially to cosmologists who are concerned with the origin of the universe. Since they approach this problem from different directions, their answers rest upon their theories regarding the origin and early history of the solar system. They report that many lines of evidence suggest an age for the earth of four to six billion years. The most widely accepted figure is about four and a half billion years, which is in accord with the geologic data. Our galaxy may be ten billion years old and the universe ten or twenty billion years or even more. Some stars visible to us today, on the other hand, are probably only a few million years old.

THE EARTH'S INTERIOR

The earth, an old and relatively inconspicuous speck in the vast expanse of space, is to us a very special part of the universe simply because we inhabit it. It is our home base. We are not likely ever to get very far from it, astronomically speaking. Naturally we wish to learn as much as we can about it.

For centuries, geologists have scratched the land surfaces of the earth. Mine shafts have been sunk to depths of a mile and a half. Wells have been bored five miles down in the search for oil and gas. But these are minute fractions of the four thousand miles between the surface and the center of the earth. In recent years, however, much has been discovered about wave motions in various substances; techniques and instruments for measuring such waves in the earth have been perfected. As a consequence, it is now possible to plumb the earth to its very center and to construct a model of its unseen interior, a model that explains the wave phenomena.

Earthquake vibrations are like messengers dispatched from a source near the earth's surface to travel far inward toward its center and return to the surface with news about the interior. Much will be said in a later chapter about earthquakes; here we will merely summarize the results of their study, insofar as they relate to the earth's interior.

The earth is not, as was formerly believed, a great spherical body of molten lava enclosed within a thin crust of cooled and therefore solid rock. Inward for a distance of about eighteen hundred miles, it is essentially solid. It is built of successive shells or layers of rock surrounding a central core which has a radius of about twenty-two hundred miles and is presumably composed largely of iron and nickel. The outer portion of this core responds to earthquake vibrations as though it were liquid. The temperature must be so high that the pressure (the weight of a column of earth material extending from the depth to the surface), although tremendous, is inadequate to force it to solidify. The central part of the core, however, appears to be solid. If so, it means that throughout a distance of about eight hundred miles outward from the earth's center the pressure is so great that the material behaves like a solid in spite of the extreme heat that must prevail there.

Surrounding the central core, with its two subdivisions, is the part of the earth's interior known somewhat ambiguously as the "mantle." This extends from a depth of about eighteen hundred miles nearly to the surface, and it therefore comprises the greater part of the earth's volume. Its substance is rocky rather than metallic and it transmits earthquake vibrations everywhere as would a solid. Within the mantle there are at least three layers or concentric zones. The outermost zone of the mantle extends upward to the bottom of the "crust," another term likely to be misleading but used to designate the outermost portion of the earth. The crust is composed of rocks that can be directly observed on land areas, or secured by means of boreholes and mining operations, or from dredging the sea floor.

The surface of contact between mantle and crust, technically a discontinuity, that is, an abrupt change in rock type or condition, is notably irregular. It was first identified in 1910 by a Yugoslav scientist named Mohorovičić and hence it is generally known as "the Moho discontinuity" or just "the moho." In general, the moho averages about five miles in depth below the bottoms of the great ocean basins and about twenty miles below sea level under the continents. In other words, the crust is much thinner beneath oceans than beneath large land areas. More than that, the moho is deeper beneath great mountain ranges than beneath extensive plains. Indeed, it now appears that the configuration of the moho is a smoothed-out mirror image of the earth's surface. Where the crust is thinnest the surface of the crust is farthest below sea level and is the deep sea floor. Where the crust has average thickness its surface is near sea level, a few hundred feet below to a few thousand feet above it. And where the crust is thickest, the mountain summits and lofty plateaus rise to altitudes of many thousand feet above the sea. This recently observed feature of the earth will receive careful scrutiny when we come to the problems of mountain-making.

Most of the rocks of the earth's crust are composed of minerals that are rich in silica and alumina. Granite is probably its most abundant rock and it runs 69–74 per cent silica and 13–15 per cent alumina. Rocks of the general type of diabase are probably the next most abundant. They are comparatively poor in silica but even so they run 48–52 per cent silica and 14–16 per cent alumina. They give to the crust a specific gravity or density that is 2.75 to 3.00 times that of water. The specific gravity of the earth as a whole is 5.56. Obviously its deeper interior must be much denser than its crust. Actually the distribution of mass (density times volume) within the earth is now fairly well known as a result of the analysis of gravitational attraction between the earth and the moon and between the earth and the man-made satellites as they orbit at varying distances from the earth and in planes at various angles to the plane of its equator. The measurements involve mathematical processes that are complicated but not too complex for a properly programmed electronic computer.

The specific gravity of the mantle increases (the rock gets "heavier") with depth from about 3.3 just below the crust to about 5.5 at eighteen hundred miles down, near the "dividing line" that separates the inner surface of the mantle from the outer surface of the core. The outer part of the core appears to have a specific gravity of about 9.5 and the density probably increases to something more than 14.0 near the earth's center.

Rocks are compressible; that is, their density increases under great pressure. Laboratory studies show what happens to rocks subjected to pressures equivalent to those that must exist at depths of forty or fifty miles within the earth. These tests show that granite could not possibly be compressed at that depth to the density of even the outermost layer in the mantle, nor could the granite transmit earthquake vibrations at the speed recorded for waves passing through that part of the earth's interior. Only rocks with silica content of less than 45 per cent, which are relatively rare in the crust, could possibly qualify as components of the mantle.

No rocky material whatever could meet the requirements of the earth's core; it must be composed of heavy metallic substances. A mixture of iron and nickel is most probable because those metals are the common com-

Halves of two cores brought up during the 1961 Mohole drilling off Guadalupe Island, Mexico, where the Pacific is more than two miles deep. At right are sediments from between 1½ and 4 feet below the ocean floor. At left are sediments found between 431 and 433.5 feet below the floor. The silty clay in each has been somewhat disturbed but the horizontal stratification can still be observed. (National Science Foundation)

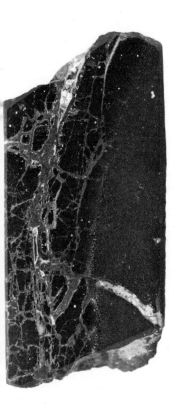

Polished cross section
of a one-inch core
brought up
by Project Mohole
from the basalt layer
about 575 feet below
the ocean floor near
Guadalupe Island, off
the coast of Mexico.
The Pacific Ocean here
is 11,700 feet deep.
(U.S. Geological
Survey)

ponents of meteorites and would have the appropriate density when compressed by the weight of the surrounding mantle. They would also be liquid at the temperature and pressure which are believed to exist at depths of eighteen hundred to thirty-two hundred miles below the earth's surface, but might well be solid nearer to the earth's center.

If it is true that there is a sphere of solid iron and nickel eight hundred miles in radius at the center of the earth, surrounded by liquid metallic material extending fourteen hundred miles farther outward to the inner surface of the mantle, one of the hitherto puzzling characteristics of the earth may be explained. The earth is surrounded by a powerful magnetic field. Consequently the freely swinging needle of a compass aligns itself in the direction of the magnetic meridians connecting the north and south magnetic poles, unless it is locally influenced by concentrations of magnetic materials. Navigation and exploration have demonstrated the importance of the compass in the life of man. At last an explanation of the fact that the earth is a huge magnet seems to be at hand. Any variation between the rate of rotation of the inner solid core and the outer solid mantle of the earth would have an electromagnetic effect somewhat similar to that of a dynamo in a power plant. Convection currents in the liquid part of the core, if not also slight changes in the rate of rotation of the mantle and crust, would presumably be able to cause such variations. The structure of the earth's interior has had subtle but far-reaching effects upon human life. This terrestrial environment has been far more favorable for us than we have yet appreciated.

It is quite likely that the conditions within the earth are unique in the solar system. Neither the Moon nor Mars can have a partially solid, partially liquid central core. The relations between their density and volume rule out that possibility. Venus, though larger, displays density-volume relations that make such a structure highly improbable. Among the objectives of the "space probes" launched by the United States and the Soviet Union is that of determining whether any of these bodies has a magnetic field of its own. It has already been discovered that the Moon has no such field, and the American "Mariner II" signaled no trace of one as it passed within twenty-two thousand miles of Venus in November 1962.

THE MOHOLE AND DEEP-SEA DRILLING

Naturally, geologists have always wanted to secure a sample of the mantle and analyze it in a laboratory. For a time, in the early nineteen-sixties, it was hoped that this desire might be realized by drilling a hole in the bottom of the sea at some place where the crust is thin enough to be penetrated by a borehole. From barges floating in water nearly a thousand feet deep, oil wells had been drilled to depths of five miles in Texas and California and to nearly a mile in the sea floor off southern California. The most difficult problem seemed to be economic: the costs would be astronomical. The operation was dubbed "Project Mohole," since the hole would have to be drilled through the Moho discontinuity.

Phase I of Project Mohole was financed by the National Science Foundation and the feasibility study was completed in 1961. Using an ocean-going barge owned by a consortium of petroleum corporations and specially equipped for drilling at sea, a hole was bored off the coast of Baja California to a depth of about 190 meters (575 feet) into the ocean bottom at a water depth of nearly three kilometers (a little less than two miles)—a world record for the time. Then Project Mohole ran into difficulties largely as a result of disagreement over long-range planning and a variety of other problems. In 1966 the Congress of the United States eliminated from the budget of the National Science Foundation any further appropriations for it.

The demise of Project Mohole was soon followed by oceanographic research of a somewhat similar nature but with a different objective and on an international, rather than solely American, basis. The new objective was to secure cores from holes drilled through the marine sediments and short distances into the underlying igneous rocks of the ocean floor at a large number of places throughout the "seven seas." By 1974, hundreds of such drill cores had been brought up from the floor of the Caribbean Sea, the Gulf of Mexico, and the Indian Ocean, as well as widely scattered parts of the Atlantic and Pacific oceans. Much of this work has been done on board the *Glomar Challenger,* a research vessel equipped with computerized positionning devices cleverly designed to keep it precisely above the drill site, regardless of winds, waves, or currents, while the drilling proceeds and drill cores are hoisted on deck. At the same time, measurements may be made of rock temperatures, heat gradients, radioactivity, and paleomagnetism—information of great significance in the development and appraisal of plate tectonics theories (see Chapter 14).

The sediments on the deep-sea floor consist of extremely fine mud and silt in which may be identified the insoluble remains of oceanic animals and plants, wind-blown dust, volcanic ash, and micrometeorites. A layer only an inch thick may represent an accumulation of a hundred thousand years or more. Most of these sediments have never been eroded or disturbed; there can be no doubt about the chronology they have recorded. The study of these samples of the ocean floor, from all parts of the globe, is opening a new chapter in the history of geology.

THE HISTORY OF THE CONTINENTS

The continuity of the geological record found in deep-sea cores is in striking contrast to the discontinuous records available on the continents. Deciphering the history of any extensive land area is like assembling an extremely complicated three-dimensional jigsaw puzzle. Careful observations made and reported by thousands of competent geologists in many parts of the world must be organized and correlated before even the broad outlines of the vast panorama are evident.

When that is done, a definite pattern is apparent, a sequence of events recurring with minor variations again and again on every continent. Earth history is fundamentally a record of the consequences of never-ending "competition" between the external and internal processes of geologic change. Internal forces that deform the crust and cause volcanic activity renew the unevenness of the earth's surface and raise portions of it above sea level. Then the external processes of erosion and deposition tend to reduce all lands ultimately to

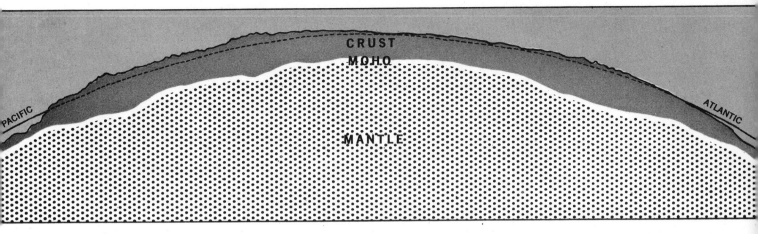

CRUST
MOHO
PACIFIC
ATLANTIC
MANTLE

an even surface at about sea level. Throughout large areas and for long intervals the external processes have been dominant, but in most regions there have been periods, in some places frequently repeated, when the internal processes have gained a temporary victory in the competition. This is fortunate for man and indeed for all the animals and plants that have ever inhabited the earth because it has created landscapes at various altitudes with diversified climates and spectacular scenery. It has made available rich resources of metallic ores and mineral fuels. It has expedited the process of biologic change, which we call organic evolution, that has produced the present assemblage of living creatures on land, in the sea, and in the air.

Each continent displays a "basement complex" composed of rocks produced before the start of Cambrian time. In some areas on each continent, these Precambrian rocks are exposed at the surface of the ground. Throughout larger areas they are now buried deep beneath younger rocks and can be sampled only in bore-holes and mine shafts. The record of events in Precambrian time is generally blurred and difficult to decipher. At many places it is evident that these ancient rocks represent the roots or cores of great mountain ranges long ago worn down to lowly plains. For the most part, one can only presume that the patterns of events during the Archeozoic and Proterozoic eras were similar to those of the better-known Paleozoic, Mesozoic and Cenozoic eras (see Geological Timetable, page 312).

We can say with certainty that every continent was repeatedly invaded during those eras by oceanic water that advanced and retreated with irregular flow and ebb. Sometimes and in some places the spreading seas reached inward from continental borders to form embayments like the modern Hudson Bay or Baltic Sea. Occasionally vast mediterranean-type seas filled long, relatively narrow, shallow trough-like depressions stretching thousands of miles across a continent. The kaleidoscopic changes in geography of land and sea included movements of the crust, some resulting in the transformation of such a trough into a mountain range. The layers of sediments, deposited in the shallow seas and changed to rock, were caught in a gigantic vise that crumpled and squeezed them into great folds and overthrust blocks. Elsewhere the flat layers of sedimentary rock were uplifted to become plateaus or were broadly arched to produce dome-like structures. Always erosion attacked the lands above sea level and deposited its products on low-lying lands or in adjacent seas. Climatic conditions changed from time to

A section through North America, showing variations in thickness of the earth's crust from the Pacific Ocean Basin at left to the Atlantic Ocean Basin at right.

25

time and extensive deserts would alternate with lush, subtropical swamps in a single region. A "moment" ago in the geological time scale, great ice sheets spread over regions where today a temperate climate is found.

Some of the details of these bygone episodes will be considered in later chapters. But this is the time to adjust to the idea of constant orderly change in all those characteristics of the earth that seem at first glance to be so stable. The life of man is so short in relation to the life of the earth that it is easy to think of mountains as "eternal"; in the history of the continents they are in fact "mists that come and go."

THE FUTURE OF THE EARTH

The most effective key to the secrets of the future is knowledge of the past. The prospect is exciting. As far ahead as scientists can see, there is nothing to suggest any astronomic or geologic catastrophe that would render the earth unfit for life as we know it. Our planet will continue to circle the sun at a goodly distance, without serious disturbance by any other heavenly body, for hundreds of millions of years to come. In the direction the solar system is going, there is no opaque cloud of cosmic dust into which the earth might plunge. There is abundant reason to expect that the sun will continue to bless the earth with its life-supporting radiant energy for at least another billion years. Climatic and geographic changes will occur in the future as in the past. But living creatures have adjusted themselves to all such changes, at least since the beginning of Cambrian time, and they will surely be able to do so in coming periods.

To be sure, now that men are able to manufacture nuclear explosives and chemical poisons, they could so pollute the atmosphere as to make the earth uninhabitable. It would be quite an undertaking, involving great expenditures of human energy and requiring considerable organization, but the requisite materials, knowledge, and skill are now available. Discounting that insane possibility, we can expect a long-continuing opportunity for mankind to use the rich resources of the bountiful earth for his own esthetic, intellectual and physical development. The question is not how much time we will have, but what we will do with it.

3

The Rocks at
the Earth's Surface

As I pointed out in the preface, some fundamental details concerning rocks and minerals are necessary to a complete understanding of geological processes. An attempt will be made to cover these in condensed form in this chapter.

The famous sculptor Gutzon Borglum selected Mount Rushmore in the Black Hills of South Dakota for his gigantic memorial to Washington, Jefferson, Lincoln, and Theodore Roosevelt because that mountain consists of a single uniform mass of granite. The stone is hard and firm because it is made up of interlocked crystals formed when the rock solidified from a molten state. It is not in layers or beds, with planes of separation that would make it more vulnerable to the attack of the weather, nor are there any other natural features to mar the monument carved on the mountainside. The hot liquid solution of mineral matter from which this granite crystallized is called "molten magma." Any rock that came directly from molten magma is an *igneous* rock.

The ancient Egyptians who carved the huge figures on cliffs overlooking the Nile River in Nubia had to work in stone belonging to another great class of rocks, the *sedimentary* rocks. The Nubian sandstone consists of sand grains that accumulated long ago as sediment, particle by particle, layer by layer, when the region was under water. Only a natural cement, chemically precipitated from that water, holds the grains together. The rock is much softer than the Rushmore granite and many times as porous. The monuments carved in it have crumbled badly, especially along the bedding planes between successive layers or beds, but they have endured for thousands of years in the dry, equable climate. They will probably be gone, however, long before the Mount Rushmore carvings, even though the latter are in a harsher climate.

The third and last major division of rocks are those that were once either igneous or sedimentary but have since been altered so fundamentally by heat, pressure, or chemical action that their original nature is now obscure. These are the *metamorphic* rocks; and they include such familiar materials as marble and slate. Some semiprecious minerals, as well as some ores of gold, copper, and iron are found in certain of them. Like many igneous rocks, they formed beneath the surface but have been exposed by erosion in such locations as the highlands of Scotland and Scandinavia and in the central parts of many mountain ranges, including the Alps, the Andes, and the Rockies.

27

South Dakota's Mount Rushmore Memorial to Presidents Washington, Jefferson, Lincoln and Theodore Roosevelt was sculptured in massive granite by Gutzon Borglum.
(Robert Clemenz)

Above right: These colossi in Nubia were carved in bedded sandstone by ancient Egyptians.
(Laurenza: UNESCO)

THE IGNEOUS ROCKS

"Igneous" comes from the Latin word for "fire" and designates the rocks that originated in the fiery-hot solutions of mineral matter called *magma*. There are two kinds. *Intrusive* igneous rocks are usually an aggregation of crystals that grew slowly as the magma cooled and solidified where it was intruded into the previously existing rocks of the earth's crust. *Extrusive* igneous rocks tend to be glassy—with little or no readily apparent crystal structure—because they were erupted as *lava* poured out on the surface or exploded in the air; they "froze" too quickly for the molecules to organize themselves in crystals. The chemical composition as well as the physical structure of igneous rocks often reveal the way they were formed, for volatile compounds are trapped in the intrusive rocks and crystallize along with more stable components. In the extrusive process the volatile substances frequently escape as gas during cooling.

INTRUSIVE IGNEOUS ROCKS

The various intrusive igneous rocks differ from each other in mineral composition and in texture. The former is determined by the chemicals in the magma, the latter by the conditions of crystallization. Thus, rocks of similar composition may display quite different textures, and rocks having similar textures may be composed of quite different aggregations of minerals. The names of the intrusive rocks normally indicate both composition and texture.

Many intrusive rocks display a granitoid texture, consisting mostly of interlocking crystals of about the same size, large enough to be visible to the unaided eye. Crystals less than about a thirty-second of an inch across create a *fine* texture; larger crystals make up a *coarse* texture. Any igneous rock with coarse granitoid texture probably crystallized in a large, deep-seated magma chamber, miles or even scores of miles long and wide. Such a magma cham-

28

ber or the great body of rock in it is called a *batholith* (from the Latin for "deep" and "rock"). The huge monolith ("single rock") of granite in Yosemite Valley known as El Capitan was carved by rivers and glaciers from a vast Sierra Nevada batholith, at least forty miles long from north to south and fifteen miles wide.

In granite itself, many of the component grains are colorless crystals of quartz, composed of *silica* (silicon dioxide). Many more are the milk-white, pink, red or light gray crystals of *feldspar,* a compound of aluminum, silicon and oxygen, with some potassium, sodium, calcium or barium. The small number of black, flaky crystals are *biotite mica,* an iron-magnesium combination. The few black, blocky crystals are *hornblende,* a compound of silicon, aluminum, iron and sometimes other elements.

Several elements combine avidly with silica to produce complex silicate molecules, and a magma with less silica than the granite magma would give no quartz crystals. Some of the non-granites with granitoid texture are the pink or gray *syenite,* the dark gray *diorite,* and the almost black *gabbro.*

Regardless of its composition, a batholithic magma has profound effects upon the roof and walls of its chamber. A vertical section through a batholith would show some of the possibilities. Because of its high temperature and active chemical ingredients, the magma may eat its way into the walls and roof of the chamber, dissolving and digesting the rocks it touches. In this way it expands the chamber outward and upward. This process is known as *stoping,* by analogy with a method used in mining operations to extend a tunnel or enlarge a stope.

Emplacement by stoping almost invariably produces great irregularities where the intrusive rock touches the rocks it has invaded. Roof pendants may extend downward into the magma chamber. Some may break off and become *inclusions,* completely surrounded by the magma. Long after all the magma has crystallized, erosion may strip away most of the roof, leaving remnants entirely surrounded by the intrusive igneous rock.

Any rock crystallized from any magma in a batholithic chamber is practically certain to display a granitoid texture. But if the magma is injected into a small chamber close to the surface or is poured out on the surface as a

A view of the various ways molten magma forms rocks. At the bottom a huge batholith has intruded into contorted schist, with roof pendants (right of center). Magma has moved up from a deep-seated chamber through four vertical conduits. The conduit at left has fed a laccolith now exposed by erosion. The next conduit has built a laccolith still covered by sedimentary rocks. In the third, some magma has spread laterally to form a sill, but most of it has erupted to build an active volcano. The fourth conduit (at right) has never reached the surface. In the distance are lava-capped mesas, buttes, a cinder cone and a volcanic neck with radiating dikes.

This recent lava flow near the summit of the Hawaiian volcano, Kilauea, of the type known as "aa" (pronounced ah-ah), is characterized by a rough surface of jagged blocks. (Hawaii Visitor's Bureau)

thick lava flow, it must cool very rapidly; time is not available for the molecules to organize into large crystals. The surface of a thick lava flow will quickly become glassy, insulating the molten material beneath so that crystals may form as it cools. The best that can be expected, however, is an interlocking network of minute crystals most of which are too small to be distinguished by the unaided eye. The texture of such a rock is described as dense, or more technically as *aphanitic* (literally "not apparent"). In terms of their component materials, the aphanitic counterpart of granite is *rhyolite,* of syenite is *latite,* and of diorite or gabbro is *diabase* or *basalt.*

Sometimes after crystallization has started in a deep chamber, the magma is pushed upward toward the surface or into some other environment where solidification is much more rapid. The resulting rock displays a combination of two textures—a sprinkling of larger crystals called *phenocrysts* (literally "crystals that show") throughout a ground mass having a fine granitoid, an aphanitic, or even a glassy texture. Such a two-stage texture is called *porphyritic,* and the rock is a *porphyry.* Porphyries are named either from the texture and composition of the ground mass, such as granite porphyry, diorite porphyry, rhyolite porphyry, or from the minerals forming phenocrysts, such as quartz porphyry, feldspar porphyry, hornblende porphyry, etc.

Sometimes magmas at intermediate and shallow depths may spread laterally between layers of sedimentary rocks. The resulting rock body, known as a *sill* (Plate 8), will be fairly flat, like one of the sedimentary beds. Quite commonly, sills are composed of basalt or diabase. Again, the intruding magma may "pile up," presumably above the pipe through which it ascends, and occupy a dome-shaped space beneath up-arched strata. The igneous rock will then be shaped like a lens. This is known as a *laccolith.* The lacco-

liths of the Henry Mountains, Utah, and of the Black Hill in the Pentlands, Scotland, are composed of porphyritic rocks.

Still other magmas may enter cracks or fissures that have considerable lateral extent and cut across or through older rocks. The rock body will then be shaped like a tablet, with its length and depth much greater than its thickness. This is known as a *dike* (Plate 6). Many dikes are composed of basalt or diabase.

EXTRUSIVE IGNEOUS ROCKS

Molten magma that reaches the earth's surface, either by ascending through long cracks or fissures or through pipelike conduits, is known as *lava,* a term applied both to the fluid material and to the solid rock formed when it cools. Lava may spread out from fissures over hundreds of square miles and may be extremely thick, as in the Deccan Plateau in India, built up by many successive flows, piled one on top of another. When lava reaches the surface through a more or less circular conduit, it commonly erupts explosively and builds circular cones and volcanoes.

The rocks thus formed are of several types. If the lava chills so rapidly that scarcely any crystals have time to develop, the rock will have a glassy texture. "Glassy" here refers to the lack of arrangement of the molecules, not to luster; glassy-textured rock may shine like glass or be as dull as unglazed porcelain. Although the rock is composed of mineral matter, there are almost no specific minerals present, and the glassy rocks cannot be named on the basis of mineral content, even though the liquids from which they

Above: Pahoehoe lava, in contrast to aa lava, has a smoother surface and is ropy or billowy. (Tad Nichols: Western Ways Features)

Above left: A close-up of jagged lava on Mount Etna in Sicily, showing the effect of shrinkage caused by the cooling of the lower surface of the solid crust of a flow after the liquid lava had drained away from beneath it. (Othmar Stemmler)

31

developed might have been rich or poor in silica or alumina, iron or calcium. Such names as *obsidian, pitchstone,* or *scoria* therefore indicate texture, luster, or structure rather than composition.

The surface of a solidified lava flow may have a wavy or rippled appearance as a result of the movements of the stiff, viscous liquid just before it "froze." Hawaiians distinguish two kinds of lava, "aa" and "pahoehoe." The former has a rough, irregular surface covered with pits made by the escape of gas bubbles. The latter has a smoother, gently undulating surface, wrinkled in places into ropelike forms.

Many lava flows are so thick that only the surface cools rapidly enough to prevent crystallization. The greater part of their mass therefore becomes a dense rock with finely crystalline or cryptocrystalline texture resembling that in a sill or dike. Basalt or diabase is commonly the result; less commonly, rhyolite. At many places basaltic lava flows display spectacular columnar jointing (Plates 9 and 10). This is the result of shrinkage in volume that occurs in all rocks during cooling. The basalt of these flows becomes solid at temperatures of several hundred degrees Centigrade. Soon thereafter it cools to less than a hundred degrees. The shrinkage tension that builds up within the entire body of rock is relieved most successfully by the development of three cracks radiating at angles of 120° from a hypothetical line perpendicular to the cooling surface. If the lines from which the cracks radiate are evenly spaced and vertical, hexagonal columns of uniform size will be produced.

Vertical sections through some lava flows show that crystallization in the lower part of a flow has produced a dense, compact rock, whereas in the upper part of the same flow the rock contains innumerable small, rounded holes called vesicles. These indicate the presence of bubbles of gas during solidification. Evidently, gas dissolved in the liquid lava tried to escape upward like the carbon dioxide in a ginger ale bottle when pressure is relieved by lifting the cap.

The potentially volatile ingredients of molten magma, which are responsible for the gas-bubble holes in many lava flows, have another consequence. Eruptions of volcanoes are due only in part to the violent formation of steam when molten magma encounters water-soaked rock or soil; much of the energy comes from the sudden expansion of gas as the magma ascends. Such explosions produce still another class of extrusive igneous rocks. These are the *pyroclastic* (fire-fragmented) rocks. Explosive eruptions may throw into the air shreds or pellets of molten lava, chunks of already consolidated erupting lava, and fragments of older rocks from the conduit walls and crater. Gobs or drops of lava still liquid enough to become rounded or spindle-shaped in flight are known as *bombs* if their diameter is greater than about an inch, and *lapilli* if somewhat smaller. Angular or irregular fragments are known as *blocks* or *cinders* and anything less than a thirty-second of an inch in diameter is *ash* or *dust.*

Ejected material of all sizes and shapes may accumulate in layers on the ground, as at Pompeii in Italy (Plate 94), or contribute to the building up of volcanic cones. Once the pieces of ejected volcanic material settle, they may weld themselves together because of their heat and plasticity, or they may be compacted as they pile up, layer upon layer. Sometimes the fragments are cemented together by the deposit of mineral matter from the water that

1. Near Tuba City, Arizona, beds containing marine fossils overlie coal seams resting on jointed sandstone. After the sand had consolidated into sandstone, this region became a vast peat bog and later a sea in which marine beds were deposited. Then came the uplifting of the region and finally the cutting of the canyon. (Paul Miller)

2. The "Beehive" (below), in the "Valley of Fire," Nevada, is an eroded remnant of conspicuously crossbedded eolian sandstone. (Mary S. Shaub)

3. The bright red sandstone in this small, weather-beaten butte (bottom) in the Navajo Indian Reservation, Arizona, is composed of sand grains bound into solid rock by films of red iron oxide. (Mary S. Shaub)

4. Iron and manganese oxides are mainly
responsible for the striking hues of the sandstone
walls in Zion National Park, Utah, and also
help cement the sand grains into solid rock.
(Robert Clemenz)

5. Stratification in some sedimentary rocks, as in the Badlands of South Dakota (above), is revealed by differences in color rather than by bedding planes. Here beds of siltstone and mudstone have been eroded by runoff from torrential rains. (Kirtley F. Mather)

6. The vertical dike (left) cutting through the alternating beds of sandstone and shale on an Arizona hillside is a basaltic igneous rock. (Tad Nichols: Western Ways Features)

7. Despite its irregular layering, the volcanic tuff (facing page) in Chiricahua National Monument, Arizona, is an igneous, not a sedimentary, rock. (Mary S. Shaub)

8. The sill of igneous rock (above) in Yellowstone Park, Wyoming, was intruded amidst a huge pile of volcanic ash erupted long before the canyon was eroded. Its uniform "palisades" result from columnar jointing. (Andreas Feininger)

9. The blocks of basalt (facing page, above) of the Giant's Causeway in North Ireland look amazingly like hand-cut stone. (Mary S. Shaub)

10. In Yellowstone the columns are vertical, but in a similar igneous rock in Iceland (right), they are gently curved or steeply inclined. (Alfred Ehrhardt Film)

11. Under extreme heat and pressure, such minerals as garnet (above, left) may result from chemical recombination of the previously existing mineral matter.
(Elbert King)

12. and 13. The gneiss (center, left) and the schist (above) display wavy bands, typical of rocks metamorphosed under great pressure at high temperatures. The schist shows a branching vein of white quartz injected into it during the final phase of metamorphism.
(John R. T. Molholm and Bradford Washburn)

14. The quartzite (left) is metamorphic rock recrystallized from a gritty sandstone in which the white granules are still visible on the surface.
(Kirtley F. Mather)

saturates them; torrential rains often accompany an eruption. If such pyroclastic igneous rock is composed mainly of volcanic ash, it is known as *tuff* (Plate 7). Tuffs deposited in beds by rain wash are generally light in weight and poorly compacted. Those in which the fragments are firmly fastened together are called *welded tuffs*. Rocks composed of angular fragments of cinders or blocks thrown out by volcanic explosions are called *volcanic breccia;* if they are composed mainly of rounded bombs, the term *volcanic agglomerate* is more appropriate.

THE SEDIMENTARY ROCKS

All sedimentary rocks have one thing in common: they are composed of small units, ranging in size from molecules up through dust particles to pebbles and large boulders, brought together and deposited on the surface of the earth's crust. In some, the components were transported by water, in others by wind or glaciers or gravity; the place of origin of some was on the land, of others in sea or lake or swamp. All the mineral matter composing them was once part of other rocks — igneous, metamorphic, or previously existing sedimentary rocks. Some of it may have passed from solution in water through the chemical processes in living plants or animals before becoming part of the rock.

Most but not all sedimentary rocks are stratified, having been put down in layers or beds, and conversely, most but not all stratified rocks are sedimentary. (Volcanic tuff or agglomerate is classed as an igneous rock even though it is often stratified.) Some sedimentary rocks are firm and fairly hard because their particles are cemented together, others because the particles were simply pressed together, and still others because they are masses of interlocked crystals that grew in cold, watery solutions.

SANDSTONE

Sandstone is one of the most frequently observed of sedimentary rocks, not because it is more abundant than any other, but because of its tendency to appear in prominent cliffs, forming the walls of canyons, the headlands along a shore, or the ledges on an otherwise smooth slope. Although diversified in origin, it is always found in beds or layers, and as a rule the *bedding planes* that separate one layer from another appear as conspicuous parallel lines on any cliff where it is exposed. Those planes are weaker than the rest of the rock and are etched more deeply by weathering and erosion. In sandstone quarries, the rocks generally break apart along such planes and yield the blocks with at least two parallel surfaces that builders often require.

Close examination of a piece of sandstone, perhaps with the help of a magnifying glass, reveals the sand grains that make up the bulk of the rock. Each is a more or less rounded particle of quartz, in many cases like grains of beach sand. The only difference is that here the grains are cemented into solid rock. The inference is clear: each bed of such a sandstone was once a layer of sand spread by currents and deposited on the bottom of a body of water. Later, the particles in the loose layers were cemented together by

41

films of mineral matter chemically precipitated from the water. Both the particles that settled down through the water and the precipitated cement are sediments.

The size of the sand grains, whether coarse, average, or fine, determines the texture of the sandstone but tells nothing about how far from shore it was deposited nor at what depth. Powerful offshore currents or undertow can carry a heavy load of coarse material far out from land. The appearance of the bedding planes sometimes reveals shallow deposition. For example, many display ripple marks identical with those found in sand in relatively shallow water. Others show rill marks or animal trails that could be made only close to shore or on a beach exposed at low tide. Evidently each bedding plane represents a change of some kind in the processes of transport and deposit of the sediment. Most commonly the change is simply a variation in the direction and velocity of currents in a restless sea or lake. The wonder is that some of the sedimentary strata are remarkably regular in thickness over large areas and that in some places many successive layers have approximately equal thickness.

In color, sandstone ranges from grayish white through shades of yellow, brown and red to very dark gray. While the individual grains play some part in coloring the rock, the cement is primarily responsible for this. The rusty yellow of many a sandstone results from the yellowish iron oxide in the calcareous cement around the grains of rust-stained quartz. This is the commonest color of sandstone, a color shared with most sandy beaches. Bright red sandstone (Plate 3) like that of the Entrada formation in New Mexico and the Bunter formation near Chester, England, gets its color from the red iron oxide staining the sand grains or present in the cement.

Neither color, texture nor bedding pattern differentiates a *marine* sandstone (one laid down in the sea) from a *lacustrine* sandstone (one laid down in a lake). Waves and currents behave in the same way in an ocean as in a large lake. Iron oxide and calcium carbonate may be dissolved in and precipitated from either fresh or salt water. There are, however, differences between the fresh-water and the salt-water faunas and floras. Certain kinds of animals and plants are confined to only one of the two environments. Thus, fossils often tell whether the rock originated in the sea or in a lake.

Fluviatile sandstones are composed of sand deposited by rivers in their channels, on floodplains or in deltas. Almost always the bedding patterns reveal the rather special origins of such sandstones. The low cliffs in the outskirts of Chester, England, plainly show the conditions under which its peculiar-looking rocks were formed. Beds two or three feet thick at one place are only a few inches thick at another place; the bedding planes curve up or down and occasionally one cuts across the bevelled edges of those beneath it. The face of the cliff is in fact a vertical section across the channels of a stream that shifted its course by "cut and fill."

Some of the most spectacular sandstones are *eolian* (wind-deposited). In arid regions the wind spreads sand across deserts or piles it in dunes. Even in humid regions, winds may build and move dunes near lake shores or inland from sandy beaches. Sometimes deep burial or submergence by advancing seas leads to cementation of eolian deposits into fairly firm rocks. A good example is the light gray sandstone in the Navajo formation exposed in many cliffs in New Mexico, Arizona and Utah. As seen in Zion National

Park, some of its beds are stained with red and yellow iron oxide (Plate 4). Such eolian sandstones are characterized by bedding patterns that announce their mode of origin. Many thin layers trend obliquely across the thicker beds to terminate abruptly against other sets of layers. Where such rocks are weathered and eroded as they are on Hoskininni Mesa, northern Arizona, their origin in the shifting sands of ancient dunes is unmistakable.

If any lingering doubt persists concerning the eolian rather than aquatic origin of a sandstone, it can sometimes be dispelled by the fossils occasionally found in the rock. The presence of land snails or other air-breathing animals and the absence of aquatic creatures are obviously significant. We can get additional information by the study under high-power microscopes of the individual grains of sand in the rock. Grains buffeted by winds and repeatedly shifted up and down the slopes of dunes become etched and "frosted" by frequent impact with one another. In contrast, the sand grains moved to and fro by waves, undertow or offshore currents are usually glassy clear because they are surrounded by a wet film rarely broken even in collisions with each other beneath the water.

CONGLOMERATE OR "PUDDINGSTONE"

Some sedimentary rocks are composed of pebbles, cobblestones or other particles too large to be called sand grains. Therefore the geologist labels them *conglomerate* rather than sandstone. In Great Britain and New England where plum pudding is frequently on the Christmas menu, this rock is often called "puddingstone" from its occasional resemblance to that delicacy. In bedding characteristics, colors and mode of consolidation, conglomerate is like sandstone; only in texture is there a notable difference. Conglomerate, too, may be either marine, lacustrine, or fluviatile in origin; it cannot be eolian for winds do not transport pebbles or cobblestones in any significant quantity.

Many marine or lacustrine conglomerates are formed of beach gravel deposited close to shore; they may blend into sandstone with distance from the ancient strandline. Occasionally a few layers of conglomerate interrupt a vertical sequence of sandstone beds. Evidently violent storms had so strengthened the offshore currents that they could transport much coarser materials. Fluviatile conglomerates are likely to be of this character because of the rapid fluctuations in volume and velocity of rivers.

Some of the most widespread conglomerates were deposited near the shores of slowly advancing seas. As the sea spread over the land in a period of submergence, its waves washed across the weathered rock debris and the soil that mantled the ground. The smaller particles were carried far out to sea by offshore currents but the larger ones were deposited close to shore where they rested directly on bedrock swept clean of other debris. As the strandline moved and the water deepened, the first layers with their pebbles, cobbles and possibly angular fragments were buried beneath layers of sand. Such a conglomerate, called a basal conglomerate, is at the base or bottom of the younger formation, resting on the older rocks of the former land surface. The lower beds of the "Old Red Sandstone," widespread in north-western England and southwestern Scotland, are almost everywhere formed of conglomerate. At many places, fragments of a light gray limestone may be identified among the cobblestones of this basal conglomerate. Thus it is clear that the limestone is an older formation than the "Old Red Sandstone."

SHALE OR "MUDSTONE"

The individual beds of sandstone are normally from two or three inches to many feet in thickness, whereas the layers of *shale* are generally only a fraction of an inch thick. Most significantly, the particles of sediment composing shale are too small to be identified by the unaided eye. The microscope shows that they are of silt or clay, with the relative dimensions indicated in the following classification of sedimentary particles by size:

(1 mm = about $1/25$ inch)	
Boulder	more than 256 mm
Cobble	64 to 256 mm
Pebble	4 to 64 mm
Granule	2 to 4 mm
Coarse sand	$1/2$ to 2 mm
Medium sand	$1/4$ to $1/2$ mm
Fine sand	$1/16$ to $1/4$ mm
Silt	$1/256$ to $1/16$ mm
Clay	less than $1/256$ mm

Many of them are thin and narrow, like minute laths; others are tiny flakes or shreds. A mass of such particles saturated with water is mud. In contrast to sandstone, shale might be called "mudstone."

Consolidation of muddy sediment to produce shale is much more a result of compaction than of cementation. The weight of overlying beds squeezes the water out and pushes the particles together so that they interlock; it resembles the way felt is made by pressing together chopped rabbit hair sprinkled over a flat surface. Such compaction reduces almost to zero the porosity and permeability of any well-hardened shale. Many firmly consolidated sandstones, in contrast, are highly porous and permeable.

Darker shale generally owes its color to considerable amounts of carbonaceous tissues of partially decayed, but not completely oxidized, animal and plant remains. Such organic debris is likely to be especially abundant in the quiet lagoons and tidal marshes along low shores in humid regions, and it is in such environments that "black shale" most commonly originates. The fine silt and clay carried by marine currents to great distances from shore is usually a much lighter gray. Beds of shale displaying brighter colors—yellows and reds, or splotches of various shades of orange and purple—are generally the result of deposit by streams in arid regions or in lakes or seas adjacent to such regions (Plate 5). Their interesting hues, like those of similarly colored sandstone, are due mainly to iron and manganese oxides.

Some sedimentary rocks may be considered as gradations between typical shale and true sandstone. For these, such terms as *sandy shale* and *shaly sandstone* prove useful. Similarly, certain sedimentary rocks may be referred

to as *pebbly sandstone* or *sandy conglomerate,* depending on the relative proportions of sand grains to particles of larger size.

The wave-cut cliffs of Staffa Island, Scotland, display basaltic lava with columnar jointing so regular that it suggests a palisade. (Annan Photo Features)

LIMESTONE

The sediments composing the foregoing rocks were brought to their resting places as particles of solid mineral matter. They were fragments, large or small, of previously existing rocks. Mineral matter may also be moved from place to place in solution, as molecules dissolved out of rocks; and several kinds of sedimentary rocks consist largely or wholly of such material. Limestone is the most abundant of these.

Limestones vary greatly in texture, hardness, color and organic content, but all are composed mainly of calcium carbonate (lime) and hence are called calcareous rocks. Carbonates dissolved in fresh or salt water may be precipitated in solid form in many ways. In the most common process, aquatic animals and plants secrete the calcareous material to construct their hard parts (shells, tests, bones, teeth, etc.). Their metabolism may also change the chemical composition of the surrounding water so as to cause biochemical precipitation. Some limestone beds, for example, consist almost wholly of fragments of seashells, more or less broken and worn by waves and currents. Others are obviously ancient reefs, built by corals and associated animals and plants. In the majority of limestones, however, the calcareous material was so pulverized before final deposit as to have been a limy mud or ooze with only an occasional recognizable fragment of organic debris. Indeed, one may find almost any variation from a *calcareous shale* through a *shaly* or *argillaceous limestone* to a pure limestone.

To determine the precise nature and origin of a limestone it is often necessary to examine it under a high-power microscope. Chalk, for example, is a soft, easily crumbled calcareous rock of light color, composed of very fine particles which are largely the skeletons of tiny single-celled marine

45

animals such as the Foraminifera. Chemical analysis also helps the geologist determine the origin of limestone and other rocks.

DOLOMITE

Such analysis indicates that many limestones consist in part of the mineral dolomite, calcium-magnesium carbonate ($[CaMg]CO_3$), rather than of calcium carbonate or calcite ($CaCO_3$) alone. When a *dolomitic* limestone is composed largely of the magnesium-bearing carbonate (as compared with calcite), the rock itself is called *dolomite*. This is one of the rare instances in which a rock and a mineral share the same name.

Dolomitic limestones and dolomites are appreciably harder than the average limestone and dissolve less easily in the slightly acid water trickling through the rocks beneath a soil cover. They are generally a darker gray than the light bluish gray of the usual non-dolomitic limestone. Their presence in the Dolomite Alps of the Tyrol adds much to the spectacular scenery of those mountains (Plate 105).

EVAPORITES

Some sedimentary rocks are in whole or in part the result of chemical precipitation caused by evaporation of water from the surface of inland seas or lagoons in arid regions, partly isolated from the open oceans. Rocks formed in this way are called *evaporites*.

The most easily identified evaporites are beds or deposits of *rock salt, gypsum,* and *anhydrite*. These are soluble in ordinary rain water and consequently they rarely remain at the surface except in arid regions. They have, however, been encountered at considerable depths in many wells drilled for oil and gas, and extensive undergound workings in many parts of the world are exploiting valuable deposits. Evidently evaporites constitute a much larger part of the earth's crust than is visible on the surface.

Thick beds of rock salt underlie a large area in Michigan, Ohio, Pennsylvania and New York, and another area almost as extensive in the vicinity of Stassfurt, Germany. In both regions the salt ($NaCl$) has been mined or otherwise recovered for more than a century. Some of the evaporites at Stassfurt contain a high percentage of other chlorides such as potash (KCl). Potassium-rich evaporites have also been discovered by exploratory boreholes in the vicinity of Carlsbad, New Mexico.

COAL AND OIL SHALE

Still another kind of sedimentary rock results from the accumulation and burial of organic material mixed with small amounts of inorganic mineral matter. *Coal seams* originated as layers of plant debris, generally in swamps or marshes such as the Great Dismal Swamp in Virginia. The peaty vegetation was buried by wind-blown or water-borne clay, silt and sand. Repeated alternation of conditions permitted the deposit of successive layers of peat

and clay. Carbonization of the layers of vegetal debris then results in coal of various grades—*sub-bituminous, bituminous,* and *anthracite*. Thus, coal seams now appear as part of a sequence of shales, shaly sandstones and sandstones in major coal fields (Plate 1).

Some fine-grained, usually dark-colored sedimentary rocks contain organic matter that yields oil when heated. If each ton of rock gives more than ten or fifteen gallons of oil, the rock is referred to as an *oil shale*. Sedimentary rocks of this kind may be either marine or lacustrine in origin. Some of them are ordinary dark brown or black shale; others, like the oil shale of West Lothian in Scotland and the Green River formation of Colorado, Utah and Wyoming, are highly calcareous with a relatively small percentage of clay minerals. In either case, the organic matter from which the oil is distilled consists largely of pollen, spores and plant fragments carried into the sedimentary basin by wind or streams. Such organic matter constitutes not more than ten per cent of the oil shales, in striking contrast to the more than fifty per cent of organic matter in low-grade coal.

The intricate crossbedding of some of the sandstones in Hoskininni Mesa, Arizona, indicates that they originated as eolian, or wind-laid, deposits of sand. (Josef Muench)

47

The third major class includes many interesting and valuable rocks that are familiar to all of us, as for instance, the various marbles, the slate used for blackboards, roofing and patios, some of the ores of gold, copper and tin, and the sources of certain gems like garnet.

All the rocks in this class were once either igneous or sedimentary, but they have been so changed by pressure, heat, or chemical action of liquids or gases that their original nature is more or less completely obscured. They are therefore known as *metamorphic rocks,* meaning simply "changed in form." The name concentrates attention upon the processes by which they evolved; these processes vary, but they all come under the general term *metamorphism.*

The intense pressure essential in producing many metamorphic rocks may be a result of mountain-making movements within the earth's crust. Sometimes layered rocks are bent or wrinkled or broken; or some of them are thrust one over another; or horizontal compression changes the shape of an entire body of rock. Such deformation entails movement within the rocks that are being compressed; movement in turn involves friction, and friction produces heat which stimulates molecular activity and causes both chemical and physical changes. Tiny crystals with the same chemical composition unite to form larger and more perfect crystals. New combinations, possible only at high temperature and under great pressure, form minerals unknown except in metamorphic rocks. Elongate crystals shift their position so that their longer dimensions become more or less parallel, at right angles to the greatest stress. As a result, the constituents of the rock are rearranged to produce the stripes and wavy bands so often observed in such rocks.

Rocks may also be subjected to considerable pressure and high temperature through burial deep below an accumulation of younger sediments. In the Gulf of Mexico off the coast of Louisiana, for example, the sediments, and sedimentary rocks beneath them, are piled layer upon layer to a thickness of at least forty thousand feet. The weight of this burden upon the lower layers, together with the temperature eight miles down in the earth's crust, are enough to start the process of metamorphism. Metamorphism in a "basement complex" and in the central part of a folded mountain range probably results from both horizontal compression and deep burial. In any event, such conditions occur throughout areas of hundreds or thousands of square miles, and the process is therefore known as *regional metamorphism.*

In contrast, great stresses and high temperatures, accompanied by chemically active liquids and gases, inevitably change the walls and roof of a magma chamber, especially if the magma is pushing upward. Consequently, many igneous rock bodies are surrounded by an "aureole" of rocks greatly altered by such *contact metamorphism.* These metamorphic rocks are important to man because they contain many high-grade metal-bearing ores. The tin ores of Cornwall, England, which have been mined for almost three thousand years and are now practically exhausted, formed during contact metamorphism when a granite batholith intruded into sedimentary rocks. The ore veins, some nearly fifty feet wide, were deposited by magmatic solutions as the granite crystallized. They extend only a few thousand feet from the granite into the adjacent slate. Similarly, the much more extensive

Many of the pebbles in this conglomerate on Bonaventure Island, Quebec, were rounded by waves before they washed up on a gravelly beach. They were long ago cemented into firm rock that has now been exposed by erosion. (Benjamin M. Shaub)

tin ores of Bolivia were deposited in a zone of contact metamorphism surrounding intruded igneous rock.

Among the changes in rocks undergoing metamorphism, recrystallization is paramount. During the early steps of solidification of an igneous rock, the crystals that form from molten magma can develop freely because they are in a liquid, whereas during recrystallization, all the new crystals are encumbered by the old minerals. Consequently the structures in metamorphic rocks generally reflect the conditions in which they developed and make it possible to trace their history.

FOLIATED METAMORPHIC ROCKS

In some metamorphic rocks, the minerals are arranged in sheets or bands differing from each other in color, composition, and sometimes in texture. These sheets are called *folia,* from the Latin word for leaf, and such rocks are said to be *foliated.* Individual folia range in thickness from a small fraction of an inch to more than a foot. Often they are wrinkled or wavy, and change in width or even pinch out completely as they are traced across an outcrop. Folia that are straight and thin generally prove to be planes of weakness along which the rock will readily split apart.

Examination of the crystals under a microscope reveals that the banding or foliation arises only in part from the segregation of minerals. It also results from the parallel arrangement of tablet-shaped, lathlike minerals within each band. Flakes of mica, if present, almost always lie approximately parallel to each other. Some of these oriented minerals may be new minerals formed during recrystallization. As they grew larger, they tended to elongate themselves at right angles to the direction of greatest pressure.

A coarsely-foliated metamorphic rock composed of the same minerals as granite is called *gneiss* (Plate 12). Its somewhat irregular stripes of mica, hornblende, and other dark minerals alternate with lighter bands of interlocking quartz and feldspar. In some gneisses the stripes curve and twist and vary so much in width that they appear to have solidified suddenly from a stiff liquid stirred by a ladle in an enormous caldron. Other gneisses suggest that a magma partially dissolved a dark rock and then injected countless thin sheets of granite. One of the more interesting of the almost infinite variety of foliation patterns consists of round or oval clusters of dark minerals, an inch or two in diameter, surrounded and sometimes connected by streaks of alternating light and dark minerals in a way that suggests the eye of an animal. It is therefore called *augen gneiss,* from the German word for eye.

In all these examples, the individual bands or folia are from an inch to a foot thick. Gneisses in which the bands are less than an inch across are said to be *schistose*. If the bands are only a small fraction of an inch wide, the rock is no longer called gneiss but *schist,* another common kind of metamorphic rock. Its thin folia usually consist of many parallel flakes and blades of such minerals as mica and the greenish chlorite. If the original rock was a sandy shale or shaly sandstone, the mica resulted from recrystallization of the fine particles of clay. An igneous rock can become schist only if it has little or no quartz but is rich in silicates that can be changed into flaky or fibrous crystals (Plate 13):

In most schists, the closely spaced folia are somewhat irregular or are slightly wrinkled. Occasionally they wrap around fairly large crystals of such minerals as garnet (Plate 11), tourmaline or staurolite. These are new minerals, formed by chemical recombination during metamorphism. The force of their crystallization was so powerful that they grew to considerable size, regardless of their surroundings. The staurolite crystals are frequently twinned so that they look like a small cross an inch or so long. The twinned crystals have such stability that in central New England and other locations they often remain unscathed even though the bedrock is deeply weathered. They are treasured as "lucky stones."

Any schist breaks more easily along its planes of foliation than across them, but it does not do so as smoothly as *slate,* a metamorphic rock resulting from the alteration of shale. The closely spaced folia of slate are generally bounded by smooth planes on which the silvery sheen of crystalline flakes of mica often catch the eye. These and other fibrous minerals, all in parallel arrangement, permit the rock to split or cleave easily into the thin slabs that make slate a valuable rock.

A variety of names are applied to other, less widespread foliated metamorphic rocks, sometimes only locally or for commercial purposes. "Soapstone" or "steatite," for example, is a talc schist, i. e. a schist composed

In the sedimentary rocks in many cliffs in the Moab Desert, Utah, firmly consolidated sandstones, as shown here, alternate with beds of sandy shale. (Fritz Goro: *Life Magazine)*

The horizontal lines in this marble quarry near Poughkeepsie, New York, result from the quarrying operation. All evidence of bedding planes disappeared when the original limestone was metamorphosed into marble. (Benjamin M. Shaub)

predominantly of talc, the mineral that, when finely pulverized, becomes talcum powder. Again, "serpentine marble" is not a true marble but a crudely foliated rock like soapstone. It is composed largely of serpentine and chlorite, formed by recrystallization of a calcareous shale, and is sold commercially as a type of marble called "verde antique."

QUARTZITE AND MARBLE

The two major types of non-foliated metamorphic rocks are *quartzite* and *marble*. Each was initially a sedimentary rock. Although the original bedding planes may be visible, neither displays foliation resulting from metamorphism because the pressure was less intense, generally speaking, than that which produces a typical gneiss or a true schist. Even so, like all other metamorphic rocks, they are recrystallized.

52

The sheer cliffs rising from the water's edge on either side of Devil's Lake in the Baraboo Hills of central Wisconsin provide an excellent exposure of quartzite in beds tilted somewhat from the horizontal. These beds, four to fifteen feet thick, are crossed at regular intervals by approximately vertical joint planes. Both kinds of plane are planes of weakness; many of the huge angular blocks in the coarse rubble mantling the lower part of the cliffs have smooth faces that evidently were once the planes of jointing or of bedding. A hand lens reveals the texture and fine lamination of a bed of sandstone. The inference is obvious: quartzite is recrystallized sandstone.

The process of recrystallization whereby sandstone becomes quartzite involves the precipitation of silica from watery solutions to replace the cements that bound together the grains of sand and to fill the pores between those grains. Having changed into a mass of interlocking crystals, the rock breaks as readily through the original grains as between them. Microscopic examination of thin slices of quartzite shows that in many instances recrystallization has "repaired" worn, buffeted and partially destroyed crystals. In some instances the original sand grain is outlined by a thin film of iron oxide that was part of the sandstone cement. Even though this film separates the new quartz from the old, the orientation of the crystal structure is now the same both inside and outside the original grain of sand. Geologic eras may have elapsed between the time of crystallization of the quartz that later became a grain of sand and the time of recrystallization of the sandstone to produce the quartzite. But the laws that govern the organization of molecules in crystals are evidently eternal and can exert their influence through such barriers as films of cement enclosing grains of sand.

Transformation of a sandstone into a quartzite requires silica-rich, watery solutions. These can do their work under moderate pressures and at temperatures found not far below the earth's surface. Hence quartzites (Plate 14) sometimes occur in places that have never undergone intensive regional or contact metamorphism. Some sandstones in undisturbed regions show incipient recrystallization and every gradation is known between a slightly quartzitic sandstone and a fully recrystallized quartzite.

The other important non-foliated metamorphic rock is *marble,* a rock name restricted by geologists to recrystallized limestone or dolomite. Its dominant mineral is calcite, which deforms readily by plastic flow even at low temperatures. Consequently, recrystallization of limestone under moderate stress generally develops the flow texture that gives some of the more colorful marbles their characteristic patterns. Any suggestion of layers in a marble quarry probably results from the plastic flow of impurities rather than from stratification.

Dolomitic marble arises not only from recrystallization of a dolomitic limestone but also from chemical reaction of ordinary limestone with magnesium-bearing gases and liquids produced by magmas in a zone of contact metamorphism. Although somewhat harder than limestone marble, it is easily cut or carved and takes on a high luster when polished.

The color and texture of either kind of marble are highly variable. The mottled and other curious patterns of many commercial marbles are due to irregular streaks or blotches of talc, chlorite, iron oxides, or other "impurities." Occasionally the fossils are not destroyed during recrystallization of a limestone and their outlines can be seen on polished surfaces; clamlike

53

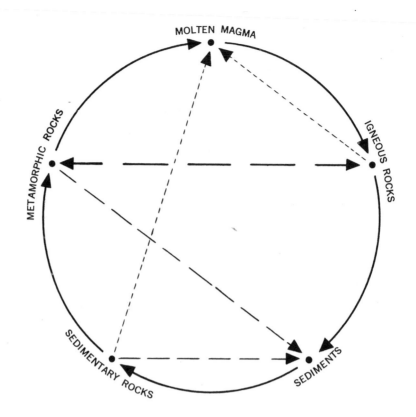

MOLTEN MAGMA

METAMORPHIC ROCKS

IGNEOUS ROCKS

SEDIMENTARY ROCKS

SEDIMENTS

The "normal" metamorphic cycle showing the change from molten magma to mineral matter and the several possible short cuts that mineral matter may take.

or snail-like shells and corals can sometimes be identified by their milky-white traceries against the gray or brown background of certain marbles widely used for interior partitions. Statuary marbles, such as the famous Carrara marble of Italy, are pure calcite, recrystallized from limestone unsullied by clay, silt, or iron oxides. Ornamental and architectural marbles are chosen for their color and for their uniformity of texture, which permits cutting into slabs or columns. Some are brightly colored by "impurities" and some are crisscrossed by white veins of calcite deposited in cracks after the main body had recrystallized.

Any rock that takes a polish is likely to be labeled "marble" for commercial purposes, but many of these are not true marble by the geologist's definition. Thus an ordinary compact limestone or dolomite in which there is not even incipient recrystallization may be quarried, cut into slabs or blocks, and polished to be sold as marble. On the other hand, a thoroughly recrystallized limestone, too friable or crumbly to be polished, would be scorned in the marble trade. Actually, nature produces every gradation from crystalline limestone through a partly recrystallized limestone to true marble.

GRANITES AS METAMORPHIC ROCKS

Recent studies have led many geologists to believe that certain granite bodies are not igneous but metamorphic, having been formed by the recrystallization and chemical change of an essentially solid mass, rather than by crystallizing from a liquid solution in a magma chamber. This process, still somewhat hypothetical, is called *granitization*.

In regions of closely folded stratified rocks, granite bodies, if present, seldom have clear-cut boundaries such as one would expect around a magma chamber. Instead, there are transitional zones from granite to such a rock as mica schist. In some places the bedding of the metamorphic rocks can be traced, though somewhat vaguely, into the granite body. For example, the near-horizontal planes that give the appearance of "sheeting" in the granite near Pt. du Van in northeastern France may well be relics of the bedding of what was originally a sandy shale.

The process of granitization can take place only under intense stress and at temperatures found at great depth in the earth's crust. It requires chemically active liquids and gases permeating the entire mass of rock. These may be exhalations from magmas far below, or they may be the water soaking downward from the surface and heated by the rocks it penetrates. Strange things happen in the secret recesses of the earth's laboratory, far beyond the direct observation of men.

THE METAMORPHIC CYCLE

The reader now has enough information about the rocks of the earth's crust to appreciate one of the great concepts of geology: the metamorphic cycle. A given particle of crustal material may "travel" completely around the cycle. Starting in a fluid, it becomes part of a firm, substantial igneous rock. Then the "destructive" phase of the cycle breaks the rock down into loose particles and soluble salts. "Construction" begins again with the consolidation into sedimentary rocks and continues into metamorphic rocks which may be as durable as the strongest igneous rocks. Pushed along by processes that cause it to sink into the earth where heat and pressure are high, the mineral matter may return to its starting point, molten magma, and begin another journey around the circle. This is a continuous operation, part of the restless, never-ending change that is a universal characteristic of the cosmos.

Many short cuts are possible. Igneous rocks may be transformed directly into metamorphic rocks. Metamorphic rocks may be broken down into sediments. Sedimentary rocks may side-step the recrystallization into metamorphic rocks and go directly into solution at a high temperature, thus becoming molten magma in short order. The variations in the metamorphic cycle are bewilderingly complex, and every outcrop of rock presents a challenge: at what point in the cycle is it, and by what route—full circle or short cut—did it reach that point?

If the concept of the metamorphic cycle is as valid as it seems to be, it is possible that some, if not all, of the molecules composing minerals and rocks have traveled around it times without number. The granite in El Capitan in Yosemite Valley, for example, crystallized in its huge magma chamber only about 150 million years ago. Since the metamorphic cycle is known to have been operating for at least twenty times that span, it is entirely possible that the molecules of that granite have gone through the metamorphic cycle many times before their latest crystallization.

4

Nature's Air Brush

Cemeteries are fascinating to those interested in specimens of the earth's crust. There is nothing morbid about this; it is simply that tombstones provide generous samples of a great variety of rocks. Their smoothly cut and often highly polished surfaces afford an unusual opportunity to study mineral components and textural patterns. They also provide a nice measure of the effects of weather on the rocks over a known period of time. Many of the weather-beaten headstones in the old "burying grounds" in New England, for example, were erected between 1630 and 1850. Some of the inscriptions are now almost illegible; corners and edges of the stones are rounded; surfaces have flaked away. In what, geologically speaking, is a very short time, exposure to the weather has produced great changes. If *weathering* can affect rocks so drastically in a century or two, what can it do in a thousand centuries?

Weathering includes both *physical disintegration* and *chemical decomposition*. In general, the former causes rocks to crumble or break into fragments; the latter causes them to decay. Although quite different, the processes operate simultaneously and it is not always possible to assign specific results to one or the other. The physical phase of weathering may dominate in some circumstances, the chemical phase in others.

WHY ROCKS DISINTEGRATE

The jumbled masses of rock fragments encountered by mountain climbers on ridge crests or barren slopes above timberline are almost exclusively the result of physical weathering. The broken slabs and loosened blocks of rock near the summits of mountains are typical of countless ranges. Because chemical action has been relatively minor, each fragment is a fair sample of the solid rock from which it was broken.

Since all minerals and rocks expand when heated and contract when cooled, changes in temperature cause rock masses to crack and crumble. At high altitudes and in dry regions the surface temperature changes radically from day to night. Every mountain climber and desert traveler knows how hot rocks become in the sunshine and how cool in the shade. The changes in temperature between night and day or between sunlight and shadow affect only the surface of the rock because mineral matter is a notoriously poor conductor of heat. Expansion and contraction of the surface rock is

15. The fantastic pinnacles in Wheeler Creek Canyon, near Crater Lake, Oregon, result from weathering of volcanic agglomerate and removal of loose particles by rainwash. The tip of many pinnacles is protected by a fragment exploded during eruptions before the lake was formed. (Josef Muench)

16. Overleaf: Nature's "air brush" accounts for the sculptured columns and cliffs in Bryce Canyon, Utah. Weathering has etched the vertical joint planes in the beds of loosely cemented, sedimentary rocks. (Emil Schulthess)

17. In Goblin Valley, southern Utah, the weathering of irregularly bedded and vertically jointed sandstone has produced such bizarre effects as "The King's Men" (facing page) standing among thousands of other "goblins." (Josef Muench)

18. The Needle's Eye (below) was produced by weathering along a vertical joint plane in the granite of the Black Hills of South Dakota. (Kirtley F. Mather)

19. Weathering along joint planes in the sandstone of Echo Cliff (top), Arizona, resembles the wholly different columnar jointing of basalt. (Andreas Feininger)

20. As a result of the weathering process known as exfoliation, great slabs have broken loose from granite cliffs (bottom) in Yosemite National Park, California. (Andreas Feininger)

22. Seen through the North Window in Arches National Monument, Utah, Turret Arch appears as a butte with a gigantic tower and keyhole. As thin layers of sandstone crumble away beneath a thicker layer, the arch grows ever larger. Wind and rain carry away the debris. (Josef Muench)

21. The dissolving action of moist air and water percolating through rock produces such pitted surfaces as those at Les Baux, southern France. (Othmar Stemmĺer)

slight, but when frequently repeated it causes the rock to crack. Other weathering agents are quick to take advantage of even the tiniest of these chinks in the rock's defenses.

Whenever rain or melting snow wets the rocks and the temperature then falls below freezing, ice forms in the cracks and pores. The motorist who has had water freeze in his car's radiator and the householder in whose home the water pipes have frozen know what happens. As water changes to ice, it increases by 9 per cent in volume. Such expansion in a tiny crevice of rock acts like a wedge driven into the rock. Each time the temperature falls below freezing, the wedge is struck another blow. The action of frost in splitting rocks apart is one of the most potent forces of weathering in many mountain ranges.

Similar effects are produced by plants whose roots expand in cracks where they have sought moisture and nourishment. A rock split by a growing tree provides a spectacular demonstration of the triumph of a living organism in its struggle with inanimate nature, but in the long run the inconspicuous activity of innumerable smaller plants, from lichens to shrubs, is the major contribution of the plant world to the disintegration of rocks.

WHY ROCKS DECAY

Exposure to the weather also causes rocks to decay. The original minerals are decomposed to form new chemical substances and some of these may be dissolved and carried away in rain water. Others may be blown away by wind or washed away as sediment in streams. Or they may remain in place and accumulate as soil. Road cuts or gully walls sometimes reveal the sequence of stages downward from the surface. First there is the topsoil, generally with a high content of organic matter; then comes the subsoil with its accumulation of clay minerals; underneath that lie partially decomposed rock fragments mixed with weathering products resting upon the parent rock, which is only slightly weathered. Chemical weathering can affect completely solid rock masses but it is much more active if the rock has been previously broken up by physical disintegration.

Rock decomposition involves four major chemical processes: *oxidation, carbonation, hydration,* and *solution*. The chemicals responsible for these processes are oxygen, carbon dioxide, and moisture, all of which are present in the atmosphere everywhere and all the time. Because of them, physical weathering is invariably accompanied or immediately followed by some form of chemical weathering, however slight. The chemical processes operate most effectively in warm, humid regions but they are also effective in some degree even in deserts and on high mountains.

When certain minerals are decomposed by combining chemically with oxygen and water, acids are formed; these in turn combine with elements in other minerals to form new compounds. Pyrite, the glittering sulfide of iron commonly known as "fool's gold," for example, may react with water and oxygen to produce sulfuric acid and iron sulfate. The sulfuric acid then attacks other minerals and produces still other sulfates.

Sulfates and carbonates are generally more soluble than oxides and silicates, but even they will not readily dissolve in cold, chemically pure

23. Where granite weathers along horizontal joint planes and is then weakened by vertical cracks, bizarre piles of slabs may result. This one, on Rough Tor, is typical of many on the tors of Devon and Cornwall, England. (Douglas P. Wilson)

24. Weathering that has followed the horizontal joints has formed the amazing "Balancing Rocks" near Salisbury in Rhodesia. (Alfred Ehrhardt Film)

The upper part of the chalk cliffs along the wave-swept shores of Les Dalls in Seine Maritime, France, is irregularly stained by organic acids draining from vegetation in farms on the flat upland. (Ray Delvert)

water. Having played so important a role in the decomposition of rocks, much water, however, is far from being chemically pure. Furthermore, even rain water contains carbon dioxide from the atmosphere, both dissolved and in the form of a weak carbonic acid. As rain seeps into the ground, it acquires additional carbon dioxide from decaying plant material. Other acids produced by plants and animals also help make the water a potent dissolving agent as it flows over the surface and percolates through rocks and soils.

Because so many different chemical compounds become available in the breakdown of rock-forming minerals, a great variety of new materials is produced during rock weathering. Among the most conspicuous are the iron oxides. Most of the iron in igneous and metamorphic rocks is combined with other elements in the silicate minerals. These commonly break down during chemical weathering and the iron produces either hematite, a red powder, or limonite, a yellow to brown powder. The rusty stains on weathered rocks and the red, yellow, buff, or dark brown colors of many soils come from mixtures of hematite and limonite.

Among the products of chemical decomposition, particles of clay are ordinarily the most abundant. Some of the components of the silicate minerals, such as feldspars, micas, and hornblende, combine with water to form the various clays. During this process, silica itself may be set free from the complex molecules of the silicates and then be removed in solution along with the carbonates and sulfates. Silica in the form of quartz, however, whether as crystals in igneous and metamorphic rocks or as sand grains in sedimentary rocks, is an insoluble substance that accumulates as small particles or grains, which are frequently blown or washed away.

66

Weathering prepares rock material for transportation by the other agents of erosion. Although the "ceaseless dripping of water will wear away the hardest stone" in time, such erosion is greatly hastened by disintegration and decomposition from exposure to the weather. Much of the sediment carried downstream in rivers comes from weathering rather than from the wearing away of rock by the stream itself. The visible sediment consists largely of disintegrated material plus the insoluble particles remaining after chemical decomposition. At the same time, the formation of soluble salts hastens the wearing away of the land. Plants take up some of the soluble products of weathering, but streams carry much of this material to lakes and seas.

Some of the more spectacular results of weathering are seen above timberline in lofty mountains where vast quantities of rock waste commonly mantle slopes or accumulate as *talus* or *scree*. These terms are synonymous and apply to the entire mass of debris; the fragmental material itself is *rock waste* or *sliderock*. The talus may form a steep slope extending outward at the base of a cliff, where its origin is immediately apparent. Loosened by weathering, the rock fragments of many different sizes have fallen or rolled from the higher crags and ledges to a position of precarious rest. The steep upper part of such a talus generally consists of larger, more angular blocks; the lower part has smaller, more rounded fragments that accumulate on a gentler slope where vegetation sometimes gains a foothold. Where the falling rocks are funneled downward through a notch or drainageway on the mountainside, a cone-shaped talus usually forms. Such talus cones are especially conspicuous in the Alps and Rocky Mountains. Adjacent cones may blend together in a continuous apron along the base of a ridge. At some places above timberline, extensive bedrock surfaces are not steep enough to permit free fall or rapid sliding of rock waste. There the loosened blocks may completely mantle the parent rock. The rubble-strewn surface is generally barren and desolate; German geologists have given it the graphic name *felsenmeer*, meaning sea of rocks. Good examples of this product of weathering may be seen in the Swiss and Bavarian Alps, the highlands of Norway, the Rockies and the Andes.

Many of the angular blocks of rock in talus or scree have one or more smooth flat sides. These are likely to represent the joint planes or bedding planes in the rock from which they were broken. In the original rock these planes provided openings for surface water and thus ice did its wedge-work there. The planes also favor maximum chemical decomposition because they retain moisture long after the rock surface has dried. Thus for most kinds of weathering they are planes of weakness.

Except for the special case of columnar jointing referred to in Chapter Three, the joint planes in most igneous rocks appear as sets of cracks or potential cracks. The cracks in any set run roughly parallel and usually nearly vertical. Different sets intersect each other at large angles. In addition, granite and similar rocks may have roughly horizontal joints that give the rock body the appearance of a stack of sheets. Where horizontal and vertical sets intersect, the rock body appears to be made of blocks placed one on the other. In metamorphic rocks, the joint planes may be either approximately vertical or at steep angles to the foliation. The joints in sedimentary rocks are likely to be more or less perpendicular to the bedding.

The origin of joints is not fully understood but it is believed that most of them are incipient partings caused by tensions within the rocks. The tensions may arise during crustal movements or when pressure is relieved as erosion reduces the overburden. They may also be a consequence of the shrinkage inevitable when water-soaked rocks lose their moisture by evaporation at the earth's surface. Changes in temperature of the rocks when erosion brings them relatively nearer the top of the earth's crust may also be involved. The great variety of joint patterns—number of sets, spacing of individual planes in a set, angles of intersection of sets with each other and with the horizontal plane, and so on—suggests a considerable diversity in their origin.

The influence of joints upon weathering explains many spectacular landscapes. The slender pinnacles known as "The Needles," in the Black Hills of South Dakota, are the product of differential weathering of a granite that displays two sets of vertical joints approximately at right angles to each other. Chemical action of water and moist air penetrates deep into the rock along the joint planes, and the slender cracks grow wider as the decomposed minerals blow or wash away. Within each vertical column the rock is still firm and some columns stand tall and free, though they are doomed to crumble away sooner or later—as their neighbors have done. The "Needle's Eye" is a spectacular example of the differential weathering process in an early stage (Plate 18).

The fantastic spires and columns surrounding the "Fairy Castle" in Bryce Canyon, Utah, were similarly sculptured by weathering along closely spaced, vertical joint planes (Plate 16). The rock here, however, is entirely different. The thick series of horizontal beds of sedimentary rock includes poorly consolidated limestone, calcareous shale, loosely cemented sandstone, and occasional layers of gravel. These vary greatly in their resistance to erosion by runoff from melting snow and summer rains. The stronger beds form the steep walls and precipitous ridges between the gullies and ravines. In them, weathering agents have produced the marvelous array of towers, minarets and needles, castles, temples and cathedrals, and the thousands of other bizarre figures that stand alone or are grouped about buttresses of the enclosing walls.

Almost everywhere vertical niches bear mute testimony to the important role of the joint planes in determining the details of the sculptures. Similarly the horizontal flutes and ribs on many spires, as well as the benches rising steplike on the faces of the castellated mesas, are unmistakably a result of the unequal resistance to weathering offered by various layers of the sedimentary rock. These vary also in color, and the haunting beauty is greatly enhanced by the intergrading or strongly contrasting bands of pastel hues of pink, orange, yellow, purple, and brown. The colors change with passing clouds, with the direction of the sun in the early morning, at noon, and in the evening, as well as when the rock is wet from a shower. They too are largely a result of weathering; most of the pigment is due to red and yellow iron oxides, lavender and purple manganese dioxide, and other products of chemical decomposition of the rocks.

Spectacular configurations of intricately weathered rocks, found at many

Vertical joint planes in horizontal layers of sandstone have influenced the weathering of a cliff above the ruins of Betatakin in Navajo National Monument, Arizona. (Andreas Feininger)

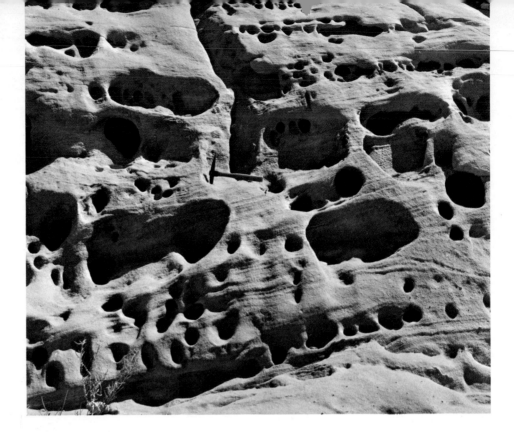

Honeycomb etching of weathered sandstone in Wadi Ram, Jordan, results largely from solution of cements that bound the sand grains together. (Taḍ Nichols: Western Ways Features)

places throughout the world, vary amazingly in the details of their shapes. The versatility of nature in producing them is matched by the imagination of the men who name them. Sharp pinnacles (Plate 15) along the crests of saw-tooth ridges are "gendarmes"; rounded columns at random on gentle slopes or in solitary grandeur in front of crenulated cliffs are "demoiselles"; pillars supporting spherical bulges of residual rock are "mushrooms," and so on. Names such as "goblin" (Plate 17) and "camelback" are suggested by resemblances to real or imaginary creatures; accidental similarities to profiles or statues of people are responsible for such as the "Old Man of the Mountain" in New Hampshire or "Queen Victoria" in Bryce Canyon.

In most cases weathering has rounded the angles of intersection between vertical joint planes and bedding planes. Many spires and needles, for example, are more or less round and all mushroom rocks have spheroidal tops. This occurs wherever the weathering is largely chemical, for the same reason that a cube of sugar becomes rounded as it dissolves in a cup of coffee. The more nearly spherical the form of a solid, the smaller is its surface area in relation to its volume and the more it resists chemical action on its surface.

This principle is apparent also in the weathering process known as *exfoliation,* the splitting of concentric thin sheets from the surface of massive rocks (Plate 20). Of all rocks, granite seems to be most susceptible to this kind of weathering. Huge curved slabs cling to the sides of the granite domes, "half domes," and "quarter domes" in Yosemite National Park, California, and are surrounded by surfaces from which other slabs have spalled away. Frazer Peak, on Stewart Island in New Zealand, and Stone Mountain, Georgia, owe their almost identical, smoothly rounded shapes to the ex-foliation of the granite composing them; on each there are incipient cracks which indicate that the process is still going on. In exfoliation, frost action combines with chemical decomposition. The feldspar and other complex minerals in the rock are hydrated by the surface moisture that percolates

70

downward for a few inches. They swell and open tiny cracks more or less parallel to the rock surface. These are enlarged by the wedgework of ice until the slabs break loose. The process is especially effective at edges and corners of joint blocks where moisture can penetrate from several directions. The separation of successive shells produces increasingly rounded forms.

If carried forward to its ultimate results, this kind of weathering leaves granite boulders such as those in the Ozark region in Missouri, in southern Arizona, in Cornwall, England, and in Rhodesia (Plates 23 and 24). Three major sets of joint planes, two vertical and one horizontal, intersect each other at approximately right angles. The roughly cubical blocks of granite, thus outlined, have been weathered to spherical shapes. Such boulders, formed in place, bear a superficial resemblance to boulders transported long distances by glaciers. Glacial markings are absent, however, and these boulders are the same kind of rock as that on which they rest.

The fondness nature displays for curves is also evident in the alcoves or shallow caves on the face of many weathered cliffs. The Great Arch in Zion National Park, Utah, is a good example of such a curve. Here the extraordinarily thick beds of Navajo sandstone consist of rounded grains of quartz bound together by calcareous cement. It is actually one of the eolian sandstones, but the fine cross-bedded laminations are not nearly so distinct as elsewhere in the vicinity. Here the weathering consists essentially of solution of the cement; the loosened sand grains either fall or are blown away. The process goes on most rapidly below the arch because water seeping to the surface concentrates there. The wide curve of the arch above the retreating face of the hollowed cliff provides the best available defense against weathering processes.

The shallow "caves" in canyon walls in Mesa Verde National Park, Colorado, had a somewhat similar origin. However, the excavation of these recesses with their vaulted roofs cannot be wholly attributed to weathering;

Landscape Arch, longest and perhaps most fragile of the natural bridges in Arches National Monument, Utah, is 261 feet long and stands 60 feet above the ground. (Josef Muench)

71

erosion by wind and by running water has been partly responsible. Such caves occur at places where less firmly consolidated and thinner beds of sandstone are overlain by stronger and thicker beds. Again the process of weathering depends largely upon solution of the cement binding the sand grains together. The concentration of moisture in certain layers and the delay before the rocks in shaded spots dry out after rain has much to do with localizing the most effective weathering. The sand grains loosened by removal of cements are quickly carried away by the strong winds in these canyons or by the runoff from infrequent rains. As usual the broad arch of the "cave" roof is the best possible defense against further weathering. The shallow caves offered shelter to the ancestors of the Pueblo Indians, the relative inaccessibility, from the canyon floor below or the mesa top above, providing an excellent defense. Between 900 and 1100 A. D., thousands of "cliff dwellers" lived in the many compact communities there. Almost certainly the rainfall was more abundant than now. The trickle of water that today seeps from the dark recess behind Spruce Tree House would be altogether inadequate to meet the needs of the several score persons who occupied the many apartments there.

If a cliff containing a weathered, arched recess or vaulted alcove happens to be the side of a narrow ridge rather than the face of a mesa or a broad upland, the recess may gradually extend clear through the ridge to form a natural bridge. Window Rock, in Utah, illustrates an early stage in this process. Many of the natural bridges in Arches National Monument and elsewhere in Utah were formed in this way (Plate 22). Other ways in which natural bridges are formed will be described in later chapters.

Thus far we have considered the rather large-scale features of landscapes in which weathering has been the dominant factor. A careful observer may also detect in almost any outcrop of bedrock the results of differential weathering on a small scale. In the vicinity of Oslo, Norway, for example, exposures of sedimentary rock consist of dark shale interbedded with layers and lens-shaped bodies of limestone. The contrast between the weathering of the shale and of the limestone is most impressive. The many small circular depressions in the surface of the limestone are due to differential solution of the easily dissolved rock. Each fresh hollow tends to be enlarged and deepened; each retains moisture long after the surrounding surface has completely dried. Once solution begins at any point, it is concentrated there; the more a hollow is deepened, the more rapidly it is enlarged. Here is another demonstration of nature's tendency toward curves rather than angles. It should also be noted that solution does not require the presence of visible water; the molecules of water vapor in the moist air react with the molecules of calcium carbonate in the limestone.

25. The erosive effect of wind-driven sand is most apparent at the base of cliffs, as in the Sahara Desert. Because of the varying hardness of the minerals in the rock, the sandblasting has created a rough surface.
(Emil Schulthess)

27. Wind sculptured the "Mushroom Rock" in Death Valley, California, from a huge block of basalt, the upper part of which was too far up to be affected by the sandblast. (Kirtley F. Mather)

26. In the Sahara Desert (top), swept by strong variable winds, lofty dunes are built up, some with rounded and others with knife-edged crests. (Emil Schulthess)

28. The dune in the foreground of this view (right) in Death Valley. California, is a barchan built by winds from the left, as indicated by the rippled sand of the horns of the crescent far to the right. (Robert Clemenz)

5

Where Dry Winds Blow

Without the atmosphere, life as we know it would be impossible and the earth, like the moon, would be a dead world, biologically and geologically. The earth's atmosphere provides the oxygen that man and almost all other organisms require. It also insulates living things from deadly solar and cosmic radiations and from countless tiny meteorites that would otherwise rain in like bullets.

An equally important life-giving attribute of the atmosphere is its circulation. If all winds stopped, cities would suffocate, the tropics would grow so hot nothing could live there, and the rest of the planet would freeze. Moisture would no longer be carried from the sea to fall as rain over the land, and the continents would turn into lifeless deserts.

From the geological point of view, the atmosphere has made its major contribution to the long history of the earth through the moisture it carries. The circulation and water-carrying capacity of the atmosphere is ultimately responsible for all erosion and construction by streams, glaciers, waves, currents and subsurface waters. But the atmosphere carries solid particles as well as water molecules, and it drops these particles, even as it does the water. In regions where the air holds little moisture, the transportation of solid matter by air becomes most important geologically and can have profound effects on the life of man. It too contributes both to erosion and to deposition.

29. Knobs and platforms of light gray, cherty limestone protruding above the rippled sands of the Sahara Desert are intricately etched by sandblast.
(Emil Schulthess)

DUST EVERYWHERE

Some of the deposits left by the wind are spectacular, like the glistening White Sands of Alamogordo in New Mexico; or enjoyable, like the clean sands above high water mark along a seashore; or awe-inspiring, like the vast expanse of dunes in parts of the Sahara; others are inconspicuous, or troublesome, or annoying. Some deposits, such as films of dust, appear everywhere; others, such as the loamlike loess, are strictly localized.

Winds are highly selective, both in picking up their loads and in carrying and dropping them. Even a gentle breeze may lift loose particles of dust from any dry surface and carry them great distances. These motes seem to hang suspended in the quietest air. The housewife needs no one to tell her that dust somehow sifts into every room and invades even bureau drawers and china cabinets. The museum curator soon learns that he must hermetically

77

seal his display cases if they are to be dust-free. It is a common saying among geologists, although no one can prove it, that every square mile of the earth's land holds at least one speck of dust from every other square mile.

Fortunately dust storms are much more localized. They occur only in arid or semiarid regions (or to leeward of such regions) where vegetation is scant and rains are infrequent. It is in "dust bowls" on treeless plains or in desert basins that fairly strong winds pick up the yellowish, angry-looking clouds and swirl them across the landscape. Actually dust is not the only annoying mineral matter in such clouds. They may include sand grains large enough to taste gritty. The coarser particles are rarely carried more than a score of miles, but the finer dust may travel many times that distance.

In addition to causing physical discomfort, dust clouds sometimes present a severe hazard to transportation by reducing visibility to only a few feet. The motorist who encounters a dust storm is well advised to pull off the road. If he cannot get the car under cover, he may find its windows and windshield pitted as though from sandblasting. Finally the particles in a dust cloud acquire electrostatic charges due to friction with air and ground, and on striking radio antennas may cause severe static.

THE GREAT DUST BOWL

But this is trivial in comparison, for example, with the disasters in the Great Plains of the United States during the years between 1933 and 1937 when extensive areas stretching northward from western Oklahoma through Kansas and Nebraska became a "dust bowl." Here the scanty rainfall has always made farming marginal. In the early 1930's many thousands of acres of prairie had been broken with a plow for the first time. Deprived of the tough, sparse grass that had hitherto protected it, the fallow or freshly seeded soil was defenseless. After two or three years of less than average rainfall, the winds began to take their toll of the parched earth. Great clouds of silt and dust swept relentlessly across the land; much of the silt was dropped in windrows on the lee side of houses, barns, and fences, burying whatever vegetation had managed to sprout there. Some of the dust, carried across the continent, caused "mud rains" and unusually brilliant sunsets as far east as the Atlantic seaboard—the small dust particles increased the scattering of short (blue) wavelengths of light and the sun appeared deep orange or red. In the dust bowl, thousands of farmers were ruined and had to take their families and migrate elsewhere. Other states were hard put to cope with the social and economic problems these migrants presented; the plight of the "Okies," as they were called, became a national concern.

There is some truth in the suggestion that the social welfare program of the federal government during the middle third of the twentieth century was prompted by the dry winds that blew across the western prairies in the 1930's. Financial assistance to the hapless farmers of the dust bowl was only a temporary expedient. The more satisfactory solution of the social problem was based upon knowledge of the ways of nature. Many of the marginal farms have been returned to pastures; trees have been planted to serve as windbreaks; most important of all, the farmers have learned to keep a cover crop on their lands during the dry season. Dry farming is now

A bluff of loess in Arizona shows an eolian deposit standing with a vertical face even though not consolidated into rock. (Benjamin M. Shaub)

fairly well understood and practiced by many farmers in dry areas; there never again need be a similar disaster in America.

DUST FROM VOLCANOES

In 1912 the violent eruption of Mount Katmai in Alaska deposited fine dust-like volcanic ash a foot thick a hundred miles to leeward. Some of the ash, blown two or three miles high in the eruption, dropped in recognizable quantities on roofs and lawns in Seattle, Washington, sixteen hundred miles away. Similar distribution of volcanic ash over large areas has been observed from scores of active volcanoes during historic eruptions. One of the latest accompanied the eruption of Genung Agung in east Bali on March 17, 1963, from which a fine brown-gray ash was carried mostly westward by a steady wind. In western Bali the deposit was five to ten inches thick and at Surabaja, 215 miles to the west on the neighboring island of Java, its thickness averaged about a half inch. Slight though that thickness seems, the ash fallout in the vicinity of Surabaja is computed to have weighed seventeen thousand tons per square mile. In the same way, the dust particles—some of them radioactive—carried to heights of several miles in the mushroom cloud of a nuclear explosion may be transported thousands of miles before they "fall out."

Under a high-power microscope, the fine volcanic ash from an explosive eruption can be seen to consist largely of minute shards of volcanic glass. Layers of it occasionally appear between beds of ordinary sedimentary rocks in ancient formations far from any known volcanoes. When fine volcanic ash accumulates on the sea floor, as it has often done, the chemical action of the salt water transforms it into a plastic, colloidal (that is, jelly-like) clay. The material is then known as *bentonite,* a name first applied to a particular clay found near Fort Benton in Wyoming. It occurs in beds ranging from less than an inch to as much as fifty feet in thickness, and is of great commercial value because of its unusual properties. The petroleum industry uses it to filter crude-oil products and in the preparation of oil-well drilling-muds. Foundries employ it to bond molding sands.

Bentonite occurs in many places throughout the world in sedimentary rocks of many ages. Deposits thick enough to be extensively mined are present in England, Germany, Yugoslavia, the Soviet Union, Algeria, Japan, Argentina, and in the United States in Wyoming, Arizona and Mississippi. Most of these are in the midst of marine sedimentary rocks of Cretaceous and Tertiary age. Interestingly enough, there is a widespread layer, only a few inches thick, in Ordovician limestone and shale in Indiana and adjacent states—the only known evidence of active volcanoes at that time in what is now the Mississippi Valley. The color of bentonite varies, but often it is yellow or yellowish-green. In weathered outcrops, it frequently displays a complicated pattern of tiny cracks, apparently the result of shrinkage as it dries out.

LOESS: A WIND-LAID DEPOSIT

At many places in the northern hemisphere and at a few places in the southern hemisphere there are deposits of loamlike material known as *loess.* It is essentially unconsolidated, yellowish-buff to gray in color, and has little or no suggestion of bedding. Although it crumbles easily between the fingers, it has a peculiar capacity to stand in vertical or even overhanging cliffs when excavated either by natural processes or by man.

The great majority of the particles in a typical loess deposit are the size of silt—larger than dust and smaller than fine sand. Most of them are tiny angular fragments or splinters of quartz and feldspar, along with some bits of calcite of similar shape. The shape of these particles explains the stability of the deposit; if they were rounded the mass would behave like loose sand. Most of the deposits of loess in Europe and North America are on uplands bordering rivers flowing out of areas covered by glaciers during the Great Ice Age.

The composition and distribution of those deposits reveal their origin. As the far-reaching ice sheets dwindled, they left the ground liberally strewn with the rock debris that had been in or on the ice. The glacial mill had ground some of that debris to powder. Crystals of feldspar from igneous or metamorphic rocks had been pulverized but not chemically altered. Crystals of calcite from the limestones in the path of the ice had similarly been broken to fragments rather than carried away in solution. Throughout a broad zone around the retreating ice-front, the cold and windy climate

Since the ripples on this sand dune in Death Valley, California, are steeper on the left side than on the right, we may assume they were formed by winds blowing from the right. (Andreas Feininger)

80

At many places, as here in Arabia, trees and shrubs killed by a migrating sand dune may be uncovered again when the dune moves on. (ARAMCO)

inhibited the growth of vegetation. Streams of water from rain and melting snow and ice rapidly eroded the abundant loose material. Dust storms swirled downwind from the near-glacial zone and from the river flood-plains veneered with glacial outwash far downstream from the glaciated areas. From the mixture of dust, silt and sand, the winds made their selection; they did not carry sand very far but they swept dust hundreds of miles. Between these regions, the particles of intermediate size came to rest and loess accumulated. Settling from the air, the angular particles interlocked with each other. Consequently, the deposit was fairly stable though never greatly compacted or bound by cement.

In Mongolia and adjacent parts of Asia, the loess generally lies on the lee side of deserts where winds blow almost constantly in one direction. There a hot desert rather than the cold desert of a near-glacial zone supplies the material.

DUNE SANDS AND SAND DUNES

Much more familiar to most people than bentonite or loess are the clean, loose sands along the inner margins of lake beaches and ocean strands or the undulating sand hills of semiarid regions and of deserts. Sand carried by winds may be spread far and wide in a thin irregular layer upon a land surface or, more commonly, in hills or ridges known as dunes. The wind seldom lifts the particles of sand more than a foot or so and usually it simply bounces or rolls them along. They come to rest when wind velocity decreases or wherever turbulent air swirls around an obstacle such as a bush, boulder or fence. Once started, the dune itself acts as an obstruction and accumulates more and more sand. Dunes are usually less than fifty feet high, but if there is an abundant supply of sand, strong winds may build them to more than one hundred feet or, rarely, over four hundred feet as in the Sahara.

Depending on such circumstances as sand supply and wind velocity and direction, dunes display a variety of shapes. Crescent-shaped dunes called *barchans* (from a Turkish word for such formations) are characteristic of certain desert regions with only small supplies of sand and with moderate winds blowing constantly in one direction (Plate 28). The horns of the crescent point downwind and the concave slope is much steeper than the convex one. The barchans in the La Joya desert in Peru are remarkable for their uniform and perfect shape and for their even distribution across the landscape. This is one of several long, narrow, completely arid deserts on the plateaus between the western Andes and the Pacific Ocean. The wind has gathered every grain of sand into dunes, leaving the surrounding desert floor carpeted with pebbles too big for it to move. Each dune marches slowly downwind as the sand is blown up its gently sloping, convex, windward face to roll down its concave lee slope, the steepness of which is determined by the "angle of rest" of the dry sand grains. The fact that each dune maintains its integrity while it moves is evidence of the constant direction and velocity of the winds.

Where similar winds have access to somewhat greater supplies of sand, they are likely to build transverse ridges at right angles to the wind direction, and every transitional form, from the perfect crescent of a barchan to scalloped ridges, may be found. Strong winds of somewhat variable direction, with a large supply of sand, pile it in longitudinal ridges parallel to the principal wind direction. If these ridges are exceptionally narrow and taper to a point, as some of them do in Arabia, they are known as *seif* (sword) dunes. Where winds are generally moderate, but variable rather than constant in direction, as in the United States and northern Europe, the dunes are characterized by irregular shapes and varied distribution, with only an occasional crescent barchan. Some of the depressions among such hills are the result of wind erosion or *deflation*. The term "blowout" is frequently applied to such depressions, although many blowouts are really formed by the building up of the surrounding ridge rather than by the deflation of the hole itself.

The primary requirement for dunes, of course, is a considerable supply of sand. Dunes are most abundant along the shores of large lakes and seas where waves and currents deposit and redeposit large quantities of sand in beaches and bars from which the wind picks it up when dry. Thus there are magnificent dunes along the southern and eastern shores of Lake Michigan, as in the Indiana Dunes State Park, where the prevailing winds are onshore, rather than offshore as on the west side of the lake. Similarly there are many splendid dunes along the Atlantic seaboard of the United States from Cape

Hatteras to Cape Cod, and especially near Provincetown, Massachusetts. At several places along the west coast of Scotland, dunes up to fifty feet high have been built by the onshore winds that pick up great quantities of sand from the North Atlantic beaches. The entire North Sea border of the Netherlands is marked by dunes, most of them now covered with vegetation; the sand came from the nearby beaches. Many of the Frisian Islands, including those farther north in Germany, consist largely of sand piled by the wind on the offshore bars deposited by waves and currents in that shallow part of the North Sea.

Next in numbers are the dunes on or near the floodplains of rivers that fluctuate greatly in volume from season to season. Here the source of the sand is the sediment borne by the river in flood and left to dry along its banks. The widely distributed sand hills of western Nebraska were formed by the winds with sand from the floodplains of the North Platte and other tributaries of the Missouri. Likewise, the dunes on the floors of Death Valley, California, and the San Luis Valley, Colorado, are composed of sand carried in by the greatly fluctuating streams that bring sediments from neighboring mountains. The San Luis Valley dunes, now set apart as the Great Sand Dunes National Monument, are predominantly on the east side, just in front of a low gap in the Sangre de Cristo Range which funnels the winds from the west.

In deserts far from river bottomlands or seashores, the sand most commonly comes from disintegrated ancient sandstones. This seems to be the origin of most of the sand in the Sahara Desert (Plate 26) although the dunes there are by no means as widely distributed as is ordinarily believed. A special case of this sort is found in the White Sands National Monument near Alamogordo, New Mexico. The dunes here are composed not of quartz but of gypsum fragments, the soft white mineral we have referred to as a common ingredient of evaporites. In humid regions the easily dissolved gypsum is ordinarily removed from the rock outcrop and carried away in solution by surface and subsurface water. Here the climate is so dry that the sedimentary rocks containing it merely disintegrate at their outcrops without much chemical change. The particles of gypsum thus become available for transport by the wind, which deals with them precisely as with quartz sand. The broken fragments of gypsum crystals are likely at first to have smooth surfaces from cleavage planes; until these are roughened by impact of grain against grain they glisten in the sunlight. The White Sands of Alamogordo are unique in the United States if not in the world for the purity of the gypsum sand and for the large area covered.

Climate is also an important factor in the formation of dunes because it influences or even controls vegetation. Thus there are many dunes in dry and desolate Death Valley, California, but few along the well-watered California seaboard in the same latitude. But even in humid regions there are many weeks in the year when sand, if present, loses its moisture and dry winds blow. The "Desert of Maine" near Freeport is a case in point. Here an area of nearly a square mile is occupied by dune sand and displays many of the characteristics of a desert, in astonishing contrast to its verdant New England surroundings. The records indicate that it was just an ordinary Maine farm until late in the nineteenth century. Then the farmer replaced the cattle in one of his pastures with a flock of sheep, to take advantage of

the burgeoning market for wool provided by the flourishing New England textile industry. He was unaware that the grass-covered, shallow topsoil concealed a thick layer of rather pure sand in a "sandplain" built by the meltwater from the retreating ice sheet of the last glacial epoch. Then came a long summer drought; the avid sheep cropped the dying grass quite to its roots. The hot summer winds scoured away the dusty unprotected topsoil and exposed little patches of dry, loose sand here and there. This was all that was necessary to release the abundant sand in the sandplain. The tiny blowouts rapidly enlarged as sand was blown over the pasture and into an adjacent orchard. In fifty years, the desert has grown from less than fifty to more than five hundred acres. To complete the story, the proprietor is probably now making more money from visitors' fees than he could hope to get from a farm of equal area. None of the available techniques for restoring fertility has been applied.

Those techniques are now well known. The clues to them were supplied by the earth itself. Everywhere, dunes tend to migrate in directions determined by the prevailing winds. Sand is driven up a windward, rippled slope to the crest of a dune and then rolls down the steeper, smoother lee side. Thus dunes march downwind, gradually invading forest or farmland, even burying villages. The speed of their migration ranges from a few feet to more than one hundred feet per year. On the Baltic coast of Germany and Poland and along the shore of the Bay of Biscay, northward from Bayonne,

Onshore winds sweep beach sand to the crest of this huge dune—more than 300 feet high—in Gironde, France. The sand then slides down the lee slope, causing the dune to migrate slowly inland. (Ray Delvert)

France, there are seventeenth-century coastal villages that were completely buried beneath advancing dunes in the eighteenth and nineteenth centuries but have come to light since then as the dunes marched farther inland. At other places, as along the shore of Lake Michigan and the North Sea coast of the Netherlands, natural vegetation has stabilized some of the dunes. The tangled roots of certain grasses and reeds bind the sand grains and prevent further movement.

The stabilizing effect of vegetation on the shifting sand has been studied at many places, notably on Presque Isle, a peninsula jutting into Lake Erie near Erie, Pennsylvania, and in the Netherlands. In a typical situation, a succession of dune ridges runs roughly parallel to the shoreline. The first ridge rising above the beach holds only scanty vegetation; most of the sand is freely shifted by the wind and the dune is a "live" one. The next ridge, away from the shore, is thickly covered with beach grasses, predominantly marram, with possibly a scattering of beach pea; the sand is largely "anchored" and the dune has begun to "die." Farther inland the successive ridges are covered with older, more permanent vegetation, including the cottonwood trees and willows of a young forest. Here the dunes are "dead." At any time, however, the live dune close to shore may creep farther inland under the stimulus of strong, constant onshore winds and override the precariously stabilized dunes in the second ridge. To prevent this and thus protect tulip fields in the lowlands behind the dune ridges, the Dutch engineers and agronomists have developed a program for stabilizing even the liveliest of live dunes. They plant the most successful varieties of beach grass in closely spaced rows over the dune and nurture them until they have gained a sure footing. The planting must be vigilantly guarded lest paths be worn across it; any break in the network of tangled roots would give the wind an opportunity to undercut the entire operation. But in the Netherlands, "Keep off the grass" means what it says and the formerly live dunes have been anchored for scores of miles along the shore of the North Sea northeastward from The Hague.

Much of the limestone of the islands of Bermuda is in reality wind-blown particles of "coral sand." Broken fragments of calcareous material, secreted by corals and other organisms, were picked up by the wind from beaches, many of them now submerged, and piled up in irregular dunes; this occurred under climatic and topographic conditions different from those that now prevail. These particles were later cemented to form the soft limestone found everywhere in this group of islands. Between the various periods in which the deposition took place, the surface was deeply weathered and the fertile soil was covered with vegetation. As can be seen in exposed cliffs and road cuts, the old soil zones form dark reddish-brown streaks in the midst of the light gray limestone. At places the original layers in the dunelike deposits of loose calcareous particles can also still be seen.

WIND AS AN ABRASIVE

Angular fragments of crystals of quartz, gypsum, or any other mineral are rounded by impact with one another as they are rolled or bounced along by the wind. When sand grains have been thus transported for a consider-

able distance, their surfaces become minutely pitted and acquire a frosted appearance. Any obstruction in the path of wind-driven sand will be abraded by the innumerable and sometimes violent blows of the sand grains. Inasmuch as particles of sand-grain size are rarely lifted more than three or four feet above the ground surface, their blows are likely to be concentrated near the base of any slender object, such as a telephone pole, rising abruptly above the ground. The wind will also climb up steep slopes athwart its path, taking along some of the tools with which it does its work. Window panes in houses overlooking beaches or exposed to sandstorms in dry regions may become frosted within a few months after the installation of clear glass. Window glass is softer than quartz and even moderate pressure or impact can scratch or etch it.

Although nature's sandblast tends to round off the angles and smooth away projecting irregularities of rocks, the varying resistance of different parts of those rocks will have a strong influence on the results. Minor variations in mineral composition and texture are responsible for many of the interesting and sometimes bizarre features often seen wherever dry winds blow. Both the behavior of the wind and the characteristics of the rocks must be taken into account in trying to explain those features.

The conspicuous and strange "Mushroom Rock" (Plate 27) alongside one

These bizarre columns, about two feet high, result from weathering of consolidated sandstone along with wind erosion at the foot of a cliff. (Tad Nichols: Western Ways Features)

Concretions on the rim of the Kharga Oasis depression in the western desert of Egypt were formed by ground water in an Eocene limestone. They remained after the crumbling rock in which they had been embedded had weathered away. Their surfaces are deeply etched by sand-blast. (Tad Nichols: Western Ways Features)

of the auto roads in Death Valley, California, is a case in point. This is a huge chunk of basalt, a fragment of a lava flow that fell from the cliff high above and is now partially buried in the debris of a talus slope. The rock is of uniform mineral composition and texture throughout. Sand has swirled around it from all directions and therefore it is everywhere rounded and smoothed. Since much more sand struck the lower part, a slender pedestal is all that is left to support the bulbous top.

More commonly, rock surfaces exposed to sandblasts vary in composition and texture. Successive layers of sedimentary rocks differ from each other and display differential resistance to abrasion. Consequently, in many arid regions they are often sculptured by the wind into weird and fantastic features. Knobs, "statues," and pillars stand out from their surroundings, defended against destruction by the harder minerals composing them, and in some places they are known as *hoodoos*.

When attempting to determine the origin of such features, it is often difficult to decide on the relative importance of weathering, wind abrasion, and rainwash. The three processes frequently work together and caution should be exercised before assigning the major role to wind sculpture. Such features as the sandstone arches previously described have sometimes been attributed solely to wind erosion, but this is incorrect. Differential weathering was predominant in their origin; the wind contributed little more than the removal of the weathered products. On the other hand, there are places where wind abrasion has sculptured holes clear through slender ridges and made "windows" in them (Plate 25).

Abrasion by wind-blown sand generally leaves recognizable autographs. A sandblasted rock (Plate 29) is ordinarily smoothed and polished, often with small grooves or flutings that indicate the direction of the prevailing blasts. Variations in hardness of different minerals or mineral clusters are shown by small-scale etching in bas-relief.

Loose pebbles or boulders in deserts often display similar markings. Because such stones have been shaped or at least marked by wind abrasion, they are known as *ventifacts,* that is to say, wind made. They are to be seen not only in the vicinity of live dunes in modern deserts but also in localities such as Cape Cod, Massachusetts, where wind abrasion was much more effective ten or fifteen thousand years ago than it is today. At that time Massachusetts was undergoing the near-glacial climate referred to on a preceding page; there was plenty of loose sand, and winds were not yet interrupted by vegetation.

Many ventifacts display wind-cut facets, generally either slightly concave or convex, but occasionally quite flat. These may bevel one or more sides of a rounded cobblestone or meet each other along fairly straight edges on stones that have acquired pyramidal shapes. Stones composed of fine-textured igneous rock, quartzite, or limestone are likely to have smooth wind-cut facets that intersect with sharp edges. Those composed of granite, gneiss, or other coarse-textured rock generally display grooves and flutings or shallow, elongated pits, which indicate the direction of the wind that sculptured the rock. On the shore of Cape Cod Bay near Stage Point there is a large granite boulder on which countless pits and grooves radiate from what must have been its windward end at the time of the sandblasting. Where the impact of wind-driven sand was direct, the grooves are an inch and a half deep; on the sides of the boulder, they are shallow flutings only a quarter of an inch in depth.

6

Running Water:
Nature's Leveler

Like many other animals, man must have an adequate supply of fresh water in order to exist. Necessarily, therefore, the major biological evolution of man took place in valleys and on plains where streams of water flowed.

Running water has exercised an equally profound influence on the social evolution of mankind. Primitive man, a food-gatherer rather than a food-grower, selected his cave, pitched his camp or built his crude huts along watercourses that supported the berries, fruits, herbs and grains he craved and that supplied the drinking water essential to his survival. Fish were speared or shot with bow and arrow in the quiet reaches of the larger streams or caught in weirs near river mouths. Shellfish were even easier to harvest. From dwelling places in rock shelters along valley walls, the prehistoric cave dwellers of the Dordogne Valley and other river basins in Europe and Asia went forth to hunt wild horses, deer and cattle, whose daily habits revolved around water holes. Stone Age hunters, like similar peoples today, undoubtedly based their hunting strategy upon their knowledge of local streams and the behavior of rivers during wet and dry seasons.

Gradually, however, primitive men became food-growers. They planted seeds, roots and saplings in bottomlands and on terraces along streams. They kept semidomesticated animals in corrals constructed near water. Certain clans or tribes in favorable regions built permanent homes and settled in farming communities. The links binding the life of man to the life of rivers became numerous and enduring. Probably the first, and certainly the most important, so-called cradle of civilization was in the Mesopotamian region, drained by streams emptying into the Persian Gulf. There, primitive agriculture and husbandry led to the earliest settlements, their locations determined largely by the regimen of running water. Long before the dawn of written history, the tillers of the ground in Asia Minor and North Africa were learning as much as they could about the behavior of rivers in order to protect their cultivated lands from flood damage and also in order to water their crops in time of drought. And men have been building dams and digging irrigation ditches ever since.

With increasing speed, civilization moved through the agricultural economy of most of man's recorded history to the industrial economy of the last two or three centuries. In many regions, the navigable rivers were for thousands of years the principal, if not the only, routes of trade and transportation. All the older cities—centers of industry and commerce—from Babylon to London and from Cairo to New York, grew up along rivers.

An aerial view of the Pasa Robles Hills in California shows a maturely dissected landscape, with many youthful valleys eroded in Tertiary shale and siltstone.
(William A. Garnett)

Caravan routes across the deserts followed the courses of intermittent streams and connected one riverside town with another. Medieval barons built their castles at strategic points along such rivers as the Rhine to collect tolls from passing vessels. Many industrial establishments were built where falls or rapids provided the power to turn the wheels.

To the observer before the age of science, running water often appeared to be capricious and uncontrollable. Much of the time, a river brought prosperity and happiness to those who dwelt along its banks. But during floods, the destructive powers that a river unleashed on helpless communities produced dire calamities. Or in periods of drought, it dwindled to a mere series of stagnant pools and ceased to sustain the people who had come to depend upon it. Quite naturally men concluded that rivers had demonic attributes and should be placed somehow in the hierarchy of gods. Thus the Ganges became a sacred river, and thus a Horatius could say, in Macaulay's words,

> "O Tiber, Father Tiber, to whom the Romans pray,
> A Roman's life, a Roman's arms, take thou in charge this day."

Where scientific habits of mind have spread, such naïve ideas have been abandoned. Rivers are now thought of as impartial natural phenomena, to be understood rather than to be feared or worshipped. Knowledge about the behavior and powers of running water makes man at least potentially able to control rivers, distributing their waters over the land according to his needs, and stabilizing their flow throughout the changing seasons.

THE LIFE CYCLE OF A RIVER

The natural history of a river and its valley resembles that of a human being or of a civilization. Normally it includes infancy, youth, maturity and old age. This sequence may be interrupted by volcanic activity or crustal movements or it may be complicated by glacial episodes. If the process of erosion proceeds to its very end, however, it will produce an almost featureless plain, approximately at sea level, traversed by a large river meandering slowly toward the sea, with a small number of large tributaries. Then, some mighty thrust which lifts up the land may rejuvenate the river system and start a new cycle of erosion in which the streams once more display characteristics of youth. Analysis of many land forms leads to the conclusion that such rejuvenation frequently occurs during maturity or early in old age—long before the work of wearing down the land to sea level has been completed.

Along the embankments of a Los Angeles coastal highway, rainwash has sculptured the poorly cemented beds of sand, gravel and clay. (Andreas Feininger)

THE INFANT STREAM

As one might expect, a typical infant stream is small and short, its valley a mere runnel, a shallow gully or a little ravine. It flows intermittently and fills its bed with running water only following a rain or when snow is melting in springtime; it will be dry much of the year (Plate 31). Its gradient, however, is so steep that it can erode the land rapidly whenever there is a

spring freshet or heavy rainfall. The tendency of such streams is to cut downward rather than sideways and to increase in length by eroding "backward" from the head, cutting farther and farther into the upland source of the runoff from rains and melting snow. These small, swift streams steal the precious topsoil from farms, gully hillsides and change fertile slopes into barren wastelands. During the latter half of the nineteenth century, such streams ruined at least four million acres of previously fertile farmlands in the United States. Most of that loss could have been avoided by contour plowing, by keeping a plant cover on the ground during the spring months instead of exposing freshly plowed earth to the heavy rains, and by clogging the heads of gullies with tree branches and dead shrubs.

THE YOUTHFUL STREAM

Sooner or later, an infant stream slices straight downward so deeply that its bed is in ground or rock continuously saturated with water in all seasons, wet or dry. Seepage from this "water table" into the stream bed thereafter gives the previously intermittent stream a permanent flow, although it may fluctuate greatly in volume from season to season. Proceeding through the various stages of its youthful career, the brook or burn or rivulet continues to cut downward and smooth out the irregularities of its sloping bed. Streams in the youthful stages of the erosion cycle are commonly characterized by falls and rapids; viewed in cross section, their valleys are V-shaped, their valley flats are only a little wider than their beds, and their valley walls rise steeply on either side (Plate 33). Such streams generally can supply water power on a small scale. The beauty of their falls, especially if they are in rocky glens or on lush green mountainsides, provides esthetic enjoyment and has inspired many a poet and artist. Brook trout and other fish, lurking in quiet reaches below their falls or above their rapids, attract fishermen. But young streams are rarely navigable, except by a canoe or small, tough boat that can "shoot the rapids," and hence they are not routes of trade or transport.

No stream can cut its bed lower than the level of its mouth. In order to carry running water a river bed must slope downward at least a little. Hence the *base level* of all erosion rises gently from sea level. The base level of erosion for a tributary stream is determined by the altitude at which it joins the larger river. All streams behave as though their most important task is to reduce their drainage areas to base level. The final accomplishment of that task marks the stage of late old age in the cycle of erosion.

MATURITY

A stream has, however, another task to perform during its transition from youth to maturity. This is the smoothing out of all inequalities in the downward slope of its bed. When falls and rapids have disappeared and the stream is flowing at a steady rate throughout its entire course, it is called a *graded stream*. The profile of its bed is a smooth curve, steeper near the source and progressively gentler downstream. The flow has approximately

the same velocity throughout the entire course, the steeper gradient of the headwaters being offset by the larger volume downstream. When a stream has thus reached grade, it has attained maturity in the erosion cycle.

It then expends more of its energy in lateral erosion than in cutting its bed downward. Swinging from side to side along its channel in ever more conspicuously meandering loops, it undercuts its banks and widens its valley flat. By this time, slope wash, gravity, and erosion by tributaries, some of which may be intermittent, have reduced the valley walls to slopes that rise gently toward distant hills and ridges. The typical cross section of a mature valley is a broad U-shape, in contrast to the V-shape of youth.

A fairly large, mature river is ideal for navigation; there are no falls or rapids, the meanders have not yet become excessively time-consuming, and there are few, if any, sandbars. Moreover, roads and railroads can be built with ease along the valley flat that borders the river. Such rivers and valleys

Seen from the air, this butte in southern Utah displays a pattern of contrasting slopes due to the varying resistance of horizontal layers of sedimentary rock. (William A. Garnett)

have long been the major routes for travel and transport in many parts of the world. The Mohawk Valley, across New York State, beckoned the pioneering Americans from the Atlantic seaboard to the Great Lakes region in the eighteenth century and today is traversed by a major railway system, the automobile traffic of the New York Thruway, and a busy barge route. The Po Valley in northern Italy, the Meuse River in France, the Sava in Yugoslavia, and the Don in the Soviet Union are equally good examples.

WHEN RIVERS GROW OLD

The transition from maturity to old age is especially long and gradual. For rivers as for men, late maturity merges almost imperceptibly into early old age. In general the meanders of a river become more sinuous and the floodplain much broader (Plate 34). Except in times of flood the stream flows slowly around the many bends in its course and frequently clogs its channel with deposits of sand and silt. When swollen with flood waters it is likely to break over its banks and inundate large areas of the valley's bottomlands. Silt and sand, deposited as floods subside, build its natural levees upward so that its floodplain commonly slopes downward away from the stream channel rather than downward toward it. On either side of the bottomlands the valley of a river in old age rises irregularly but gently toward low rounded hills at the rim of its extensive drainage basin.

Typically such a stream flows through an undulating plain with few hills and no mountains. An extensive land surface thus worn down by running water until it is nearly everywhere close to the base level of erosion is known as a *peneplain,* that is, "almost a plain." If conspicuous elevations remain standing here and there above a peneplain, they are called *monadnocks,* from Mt. Monadnock, which stands high above the rounded ridges in the southern part of New Hampshire. Ordinarily monadnocks are composed of more resistant rock than that in the surrounding region. Some of them, however, may be merely the hills that were most remote from the courses of the larger rivers and the major tributaries during the development of the peneplain.

Rivers in old age pose many problems for human beings. They are generally large, both in length and width, and hence they invite navigation. But such rivers have bad habits that make difficulties for vessels having a draft of more than a few feet. They build sandbars at unexpected places; their main channels shift from place to place almost overnight. Charts showing navigable routes and depths of water can scarcely be drafted fast enough to keep up with their vagaries; only constant dredging can maintain open waterways. The result is that all navigation, other than by barges with very shallow draft, may be halted during low-water stages. Fertile bottomlands will, moreover, be inundated when the river overflows or breaches its levees.

The Mississippi River, downstream from St. Louis, is in old age. Back in the seventeenth century, such early explorers as Joliet, Marquette, and La Salle used it as the main route for their journeys through the interior of the North American continent. Their canoes, of extremely shallow draft, carried them easily and swiftly downstream and more slowly and laboriously

These potholes in the bed of the Vaucluse River in southern France were eroded by pebbles and cobbles (see dry pothole at right) swirled around in eddies during stages of high water.
(Othmar Stemmler)

upstream. It made no difference whether the water was one foot or twenty feet deep. With the coming of steamboats and heavily laden barges in the nineteenth century, however, fairly deep channels were necessary and the behavior of a river in old age became critically important. The exploits of river pilots in the face of such difficulties have been recounted by Mark Twain and other writers in fascinating detail. The swift and treacherous currents of the river in spate were no less dangerous than the bars and shoals that appeared when the water was low. In recent years, hundreds of storage dams in the upper reaches of the Mississippi and its major tributaries have done much to equalize the flow in all seasons.

The task of controlling rivers started in Europe long before the construction of these dams on the Mississippi, but it has been carried forward with adequate tools and technical information only during the past century. In general the larger rivers of eastern Europe, such as the Volga and the Don, are in old age throughout the greater part of their length. The Don, which flows from headwaters a hundred miles south of Moscow to empty into the Sea of Azov, the northeast extension of the Black Sea, is more than a thousand miles long and drains an exceptionally fertile area of more than 160,000 square miles. Since the middle of the eighteenth century, it has produced several disastrous floods of extraordinary magnitude but the construction of storage dams has now done much to stabilize its flow. The Volga is the largest river in Europe, 2300 miles in length from its source on the Valdai plateau to its mouth on the north shore of the Caspian Sea, with a drainage basin covering more than half a million square miles. It has long been a major route for water transport but its navigation suffers all the ills of a

97

stream in old age. The Upper Volga Dam and Rybinsk Reservoir contribute much toward its stabilization, and other storage dams, completed or under construction, are bringing the river under effective control. Improvements to navigation are especially important in taking advantage of the unusual geographic relations between the Volga and the Don. Near Volgagrad (formerly Stalingrad) the two rivers lie only forty-five miles apart and because both are in old age, the area between them is very low. This has made it easy to construct a canal to carry shipping from one river to another. The completion of the canal in 1951 made direct water transport available from the Caspian Sea to the Don Valley and from the Black Sea to the Volga Valley.

The Danube, Europe's second largest river, is also in old age throughout much of its course but the topography of its valley is extremely complex. From its source in the Black Forest in southwest Germany downstream nearly to Ulm, it is a youthful stream or, at most, in early maturity. Below Ulm it flows across weak, poorly consolidated sedimentary rocks past Regensburg and on toward Passau. Here it has the characteristics of a stream in early old age, with a meandering course across a broad undulating plain. Taking advantage of this subdued topography, the much-used Ludwigs Canal connects the Danube with the Main River, one of the larger tributaries of the Rhine, thus providing a shipping route from the Danube Valley with its highly developed agricultural and mineral resources and its numerous industrial centers, to the Rhine Valley with the well-integrated complex of modern industries that make western Germany a leading technological nation. Such industrial development depends upon efficient transportation of raw materials and finished products. Since water transport is usually most economical for bulky goods, the network of navigable rivers and canals in the European "heartland" is basic to the prosperity of the people. The Ludwigs Canal is a prime example of the adjustment of mankind to rivers in an advanced stage in their cycle of erosion.

Approaching Passau and continuing throughout two-thirds of its course across Austria, the Danube is crowded between the uplands of the Bohemian Forest to the north and the outposts of the Austrian Alps to the south. Here

Below right: Diagrammatic view of the relation of the Niagara Escarpment, which cuts diagonally across the landscape, to the underlying sedimentary rocks. When Lake Erie (upper right) began to discharge down Niagara River into Lake Ontario (lower left), it created Niagara Falls where the river plunged over the escarpment. The falls have since retreated to their present position, forming Niagara Gorge.

Below: A diagrammatic section through Niagara Falls, showing the resistant limestone cliff over which the water pours and the undercut cliff where the weaker shale, sandstone, and limestone beds are worn back by turbulence in the plunge pool.

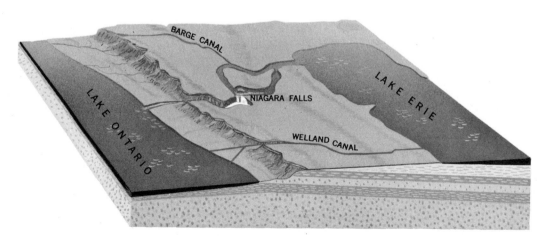

the resistance of the crystalline igneous and metamorphic rocks across which it flows greatly retards the erosion of the land. The river hurries through several granite gorges that alternate with reaches of quiet water. In this part of its long course, the Danube displays the typical characteristics of a stream in late youth. Much has been done to make the rapids passable and otherwise to improve navigation, but little use has thus far been made of the potential hydroelectric power offered by such youthful stretches in the large river.

Before it reaches Vienna, the Danube emerges from the confines of those resistant rocks and flows across the Vienna Plain and onward past Bratislava across the Little Alföld of western Hungary. There, in the weak, poorly consolidated rocks, it once more displays the characteristics of old age. Low, undulating plains stretch far on either side of the meandering river to provide fertile land for the generally prosperous agricultural economy.

Farther downstream, however, the Danube encounters additional obstacles. Resistant rocks responsible for the hills and ridges of the Bakony Forest–Matra region lie directly athwart its course and explain the scenic Danube Gorge above Budapest. The youthful character of the river and its valley here has had a profound influence on human history. Romans long ago built Aquincum as a fort to control access to the pass afforded by the swift but navigable river in the gorge. Later the Germans built Buda on the right bank of the Danube three or four miles below for the same purpose. Flowing steadily beneath the several bridges uniting the two parts of the modern twin city of Budapest, the river continues southward across the almost flat and very fertile plain known as the Great Alföld into the northeast corner of Yugoslavia. Once more it presents the characteristics of old age as it meanders in frequently changing channels across the weak rocks.

But it is not yet freed from the restraints of the complex geologic structures of this part of Europe. Turning eastward near Belgrade, it encounters the closely folded sedimentary rocks and igneous massifs of the Transylvanian Alps, the southwestward extension of the Carpathian Mountains. Probably these mountains were thrust up across the course of the Danube long ago and the river held to its route by cutting its channel downward as rapidly as the land arose. If that is true, the Danube might be called an *antecedent stream,* as a compliment to its erosive vigor. Whether or not this is the correct explanation, the river flows along the boundary between Yugoslavia and Romania in a youthful valley, which at such places as the famous "Iron Gate" is a truly spectacular gorge or canyon.

This is the last obstacle the river must overcome. Emerging from the Transylvanian Alps, the Danube quickly regains the characteristics of an elderly river and traverses the southern margin of the Wallachian Plain to empty, at long last, into the Black Sea. Here its wide floodplain is subject to seasonal overflow; its meandering channel shifts position on the slightest excuse; its waters separate into unexpected distributaries, especially where they cross the vast mud delta as they approach the sea. At many places the plain is too marshy for agriculture but away from the river it yields valuable crops of wheat and corn. Much is being done to improve navigation on the Danube and its major tributaries, and to develop the considerable potential of water power in the youthful segment of the river in Transylvania. Rarely is man so challenged by a river like the Danube, traversing half a continent and displaying such diversified characteristics in various parts of its course.

The Nile, below Aswân, is also in old age. The floodplain, which averages only about ten miles in width, extends for seven hundred miles downstream and includes much of the agricultural land of Egypt. Here too, the problem of flood control is ancient and stubborn. It is, moreover, combined with the need to get more water at higher levels for irrigation of the arid deserts that stretch far away on either side of the river's fertile bottomlands. The high dam now under construction above Aswân should prove a boon to the fellaheen of Egypt.

In general, dams in river basins where the master streams are in old age throughout much of their courses serve a multiple purpose. They improve navigation, reduce the danger of floods, supply hydroelectric power, store water for irrigation, and, as a bonus, the reservoirs provide welcome recreation facilities in regions that have no natural lakes. These several purposes, however, are not always compatible. For maximum production of hydroelectricity the reservoir should always be full or nearly full of water; for maximum flood control it should be empty or nearly empty immediately before the season of greatest danger. By and large, the function of power development is best served by dams across youthful streams or those in the earlier stages of maturity, whereas flood control and improved navigation should have priority in the design of dams for rivers in old age. Most important, the entire drainage basin of a major river system must be taken into consideration as it has been in the operations of the Tennessee Valley Authority. Adequate knowledge of the habits of the particular rivers must be gathered; rivers have almost as much and as varied individuality as human beings.

The artificial boundary lines between states and nations are often handicaps and sometimes insuperable barriers to man's task of making rivers his faithful servants. The prevention of floods that would destroy fertile farmlands in Louisiana, for example, can best be accomplished by the construction of storage dams in a dozen other states far to the northeast and northwest. Painful experiences have driven home the truth that building artificial levees on top of natural levees is like the labor of Sisyphus. As the spring floods subside, a layer of silt is deposited in the stream channel between the levees, thereby reducing the volume of water they can retain; the next year the levees must be raised still higher. In time the bed of the river is far above the level of the floodplain and disaster is well-nigh inevitable. Stabilizing the flow throughout the year by storage dams in the headwaters keeps to a minimum the silting up of channels downstream.

There is a sense in which rivers are "respecters of persons," rather than completely neutral toward mankind. Certainly they are amenable to control by persons of intelligence, ingenuity and skill who are willing and able to cooperate with each other. Only to such people do rivers bring the beneficence of well-watered, fertile farmlands, low-cost and year-round transportation facilities, and abundant hydroelectric energy.

30. At low water, the Colorado River makes a meander curve at Toroweap Point west of Grand Canyon National Park. While this "plateau country" was being lifted up during the last million years, the river cut its way down more than 5000 feet through the sedimentary rocks. (David Muench)

THE RIVER'S TOOLS

Rivers do much of the work of wearing away the land merely by transporting mineral matter, either as solid particles or in solution. Rock debris produced

31. An intermittent tributary (left)—shown during the dry season—of the Colorado River in the Glen Canyon region, Utah, has cut an extraordinarily narrow canyon, a striking example of a stream in early youth in the erosion cycle. (Tad Nichols: Western Ways Features)

33. Oneonta Gorge (right), Oregon, is a deep cleft in basaltic lava eroded by a foaming brook, youthful in behavior but with a year-round flow. (Josef Muench)

32. Although the walls of Canyon de Chelly, Arizona, are precipitous, its broad, flat floor indicates that the stream eroding it is now advancing from youth to maturity in the erosion cycle. (William A. Garnett)

34. In its meandering across flatlands in Alaska, a small creek (left) displays the characteristics of old age in the erosion cycle. (Steve McCutcheon)

35. The "Devil's Throat" (below, left)—the local name for the point where the gorge is narrowest—is in the left background in this view of the great Iguazu Falls from the Brazilian side. (Weldon King)

36. In Rhodesia, the Zambezi River plunges over Victoria Falls (right) into an awesome chasm eroded in a volcanic plateau where the lava has been shattered by movements of the earth's crust. (Emil Schulthess)

37. Overleaf: The slopes leading up from the flat floor of Death Valley, California, to Zabriskie Point are dissected by many intermittent streams eroding easily removed materials. (Robert Clemenz)

38. The spires, buttes and mesas in semiarid Monument Valley (left), Arizona, contrast strikingly with the usual landforms of well-watered regions. They have been sculptured from a jointed sandstone that rests on thinner "red beds," beneath which is the crossbedding (in the foreground) of a wind-laid sandstone. (Josef Muench)

39. The jagged pinnacles known as the Teeth Points (above, right), twenty miles southwest of Oraibi, Arizona, result from the weathering of layers of sandstone having prominent vertical fault planes. (William Neil Smith: Western Ways Features)

40. The Nonezonshi Arch (right), or Rainbow Bridge, Utah, was carved in massive sandstone by Bridge Creek. The creek used to flow around the buttress at the right before it cut through the narrow neck of an incised meander. (Western Ways Features)

41. The magnificent loop in the canyon
of the San Juan River at "The Goosenecks"
in Utah was cut by this river after it was
rejuvenated by a vast uplifting of the region.
(Josef Muench)

Running Water:
Nature's
Builder

42. Fed by silt-laden meltwater from glaciers and snowfields on the flanks of Galdhopiggen and Glittertind, the Bovra River is here building an extensive delta in a narrow Norwegian lake known as Ottavann.
(Kirtley F. Mather)

43. Stream deposits fill up the estuaries along the shore of Cook Inlet, Alaska.
(Steve McCutcheon)

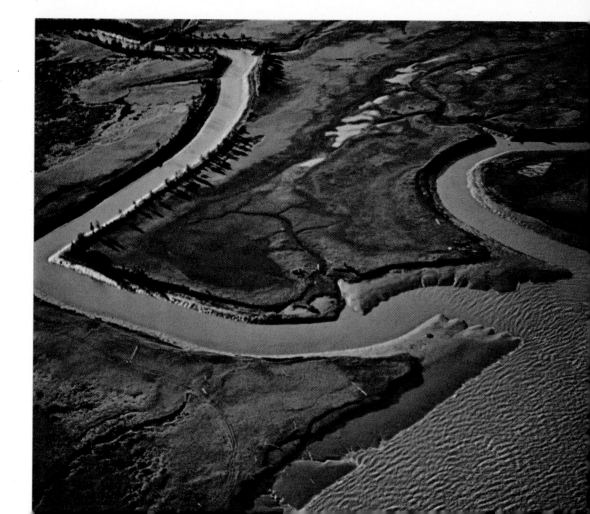

by weathering falls into streams or is blown in by the wind, or is washed in by rain and the sheets of water that flow downward over slopes during and after heavy rains. The river carries fine particles—the size of dust or clay or silt—in suspension; in the generally turbulent flow of the river, such small particles are thrust upward by rising currents before they can settle to the channel floor. Larger particles, such as sand grains, pebbles, cobblestones and boulders, are slid, rolled or bounced along the stream bed. The capacity to transport debris depends of course on stream velocity and volume. A slight change in the velocity of the flow from place to place along a stream bed or in the volume of water at any one place from time to time produces a great change in the stream's transporting ability. Hence the particles of debris it is carrying seldom complete their journey in a continuous voyage but make many stopovers en route.

Hence also, flowing water is highly selective in regard to the materials it transports and deposits. The clear, clean water of a swift mountain brook dashes along over gravel and boulders; it quickly removes the small amount of fine particles available from day to day. The large, relatively slower river in the lowlands is muddy; there is more fine material always available along its banks and in its bed than its velocity permits it to carry away, even though its volume is great.

Perhaps you have had the joy of camping for the night near a swiftly flowing stream of clear, cold water in a mountain valley somewhere beyond the reach of superhighways or railroad tracks. If so, you may have noted in the stillness the occasional low rumbling sound of pebbles and boulders rolling over each other and shifting position beneath the rippling water. This is the audible evidence of the process that rounds off angular fragments of rock debris during transport. The visual demonstration was there too. The farther downstream you go from the sources of angular weathered debris, such as rocky cliffs or talus piles, the more rounded are the stones in the stream bed.

The solid particles are the tools the rivers use to abrade their beds. Without them, running water is almost powerless to erode unweathered bedrock; with them, youthful rivers often cut narrow steep-walled gorges and reach the local base level of erosion, developing the smooth profile of a graded stream. Many spectacular chasms are essentially the result of this kind of erosion. Sheer walls of rock rise abruptly on either side of the stream bed, in some places to heights that are greater than the width of the chasm between them. Smoothly cut, concave surfaces are sometimes apparent where the eddying stream pelted the sides of its channel with the rock fragments it was swirling downstream. In some instances the stream at the bottom of the gorge cascades from one rock bench to another or from one pothole to another along its course. Occasionally partially rounded fragments of rock caught in an eddy are seen grinding away at the bottom of a pothole. Under favorable conditions even a small stream can rapidly deepen its bed by abrasion to produce spectacular results.

Streams flowing over limestone also erode their beds by dissolving away the soluble mineral matter. Both the chemical process of corrosion and the physical process of abrasion operate simultaneously. The former produces conspicuous results only in limestone regions where little sand or gravel travels along the stream bed and abrasion is relatively slow. Under those

44. In this part of its course in Wyoming, the Yellowstone River is in late maturity in the erosion cycle and is undercutting its banks on the outside of each meander curve while depositing sand and gravel on the inside. (Andreas Feininger)

113

conditions the stream bed is likely to be a succession of solution channels which are the result of corrosion rather than abrasion. The position of these channels is strongly influenced by the joint systems in the limestone.

FALLS THAT MOVE BACKWARD

Rocks differ greatly in their resistance to erosion by running water. Some are strong and their erosion is a long, slow process. Others are weak; they yield readily and are worn away more rapidly. Among the many factors that determine the strength or weakness of a rock body are such things as hardness and chemical stability of component minerals, porosity and permeability, degree of compaction (especially for shale), solubility of cements (especially for sandstone), spacing of joint planes, and thickness of layers. The best that the casual observer can ordinarily do is to reason backward: rocks obviously more resistant to erosion than their neighbors must be the stronger. Then more careful scrutiny will often enable him to make an informed guess as to why they are not being worn away more rapidly.

Many, but by no means all spectacular waterfalls are due to variations in the resistance of rocks to erosion by the streams that plunge over them. Consider Niagara Falls: aerial views show the striking contrast between the upper portion of Niagara River and the part below the falls where, in the steep-walled gorge, one can see the outcropping ledges of horizontal beds of limestone and shale that underlie the surface of this part of New York State. Looking at the American and Horseshoe Falls from the Canadian side of the river, one notes the thick beds of gray limestone in the upper part of the precipice and the thin dark layers of shale in the lower part, concealed in places by weathered debris. The limestone is strong and resistant, the shale weak and easily eroded. The backwash of the plunging cataract erodes the shale and undercuts the cliff. When enough shale has been washed away, the overlying limestone loses its support and great blocks of it come crashing down. Some of them can be seen at the foot of the American Falls, but beneath the Horseshoe Falls they are concealed in the deep "plunge pool."

Niagara Falls first came into existence twelve to fifteen thousand years ago when western New York and southern Ontario were laid bare by the retreating ice sheet just before the close of the Great Ice Age. At that time, the overflowing water in the Lake Erie basin made its way northward toward Lake Ontario. The Niagara escarpment separates the Erie Upland from the lower lands around Lake Ontario. The river plunged over that cliff near Lewiston, New York (seven miles below the present falls), and the roaring cataract was born.

Then as now, the strong limestone protected the brink of the precipice and the weak underlying shale yielded to the backwash in the plunge pool at its foot. Had the rocks been of uniform resistance, either all weak or all strong, the steep, abrupt change in the stream gradient would have been smoothed out—first a waterfall, then a cascade, and later a series of rapids—as the watercourse developed the normal profile of a graded stream. It is the unequal resistance of the nearly horizontal beds of limestone and shale that explains why the falls receded a full seven miles upstream to their present position without any significant change in the steepness of

the declivity over which the water plunges. Obviously this retreat of the falls produced Niagara Gorge and gave that portion of the river its youthful appearance.

It should be noted, however, that the relation of the river to the tools with which it works was also a factor in causing this extraordinary example of receding waterfalls. The water of Niagara River as it flows outward from Lake Erie is exceptionally clear and clean. Debris transported into Lake Erie settles to its bottom; no particles of mineral matter large enough to be effective in abrading the bed of the river are carried through the outlet of the lake. The clean blue water rushing so swiftly over the brink of the falls past Prospect Point has no tools with which to wear away the limestone. In striking contrast, the turbulent water in the plunge pool picks up angular fragments of that same rock, fallen from the undercut cliff, and hurls them violently against the weak shale. Rarely does running water have such effective tools to use so energetically. Rock structure and stream activity have worked together like a well-coordinated team to produce the awe-inspiring result.

The crest of the curving Horseshoe Falls retreated upstream about 350 feet between 1842 and 1942. At that rate it would have required only about ten

In the Grand Canyon, the Colorado River is deepening an inner gorge where resistant Precambrian rocks maintain the youthful appearance of the V-shaped chasm. Above the walls of schist and granite is the Tonto Platform, floored with beds of Cambrian sandstone. Nearer the plateau surface are buttes, "thrones," and "temples" showing sedimentary strata ranging in age from Cambrian to Pennsylvanian. (Spence Air Photo)

115

thousand years for the falls to recede from their original position near Lewiston to the present location, seven miles upstream. The problem of measuring the antiquity of the falls in years is not, however, so simple as that. The volume of water overflowing from Lake Erie to form Niagara River has not always been the same as it is now. There was, for example, a time during which the three upper Great Lakes overflowed eastward across the Ontario "peninsula" in a river that emptied into Lake Ontario east of Toronto instead of southward into the Detroit River as they do today. The change occurred when the northeast part of the North American continent moved upward after the removal of the great weight of the ice sheet. Before that, the volume of Niagara River was far less than now and the rate of erosion under the falls must have been much slower. Presumably too, the falls were not so wide as now; this would explain the relatively narrow portion of the gorge below the whirlpool rapids. There were other complications also in the history of the Great Lakes and therefore of Niagara River (see Chapter Nine). Putting them all together, geologists used to say that the total recession of Niagara Falls required twenty-five to thirty thousand years. Since Carbon 14 radioactive time measurement has become available, fairly precise dates have been determined for several positions of the ice front in the Great Lakes region; these are the basis for the statement that the falls are twelve to fifteen thousand years old.

With this history of Niagara Falls in mind, it is natural to raise the question of their future. If the river were left free to carry on its normal activities unhampered by the work of man, they would continue to recede upstream until Lake Erie was completely drained, perhaps twenty or thirty thousand years from now. But men have interfered with nature in the last fifty years and they will probably do so even more in the future. Already about one-fifth of the water of Niagara River has been diverted into penstocks to bypass the falls and produce hydroelectricity. More than twelve billion kilowatt hours of electricity per year are now being generated on the Canadian side of the river. When additional generating capacity on the American side, now under construction, is completed, more than one and a half million of the six million horsepower of energy formerly running wild will have been harnessed. This diversion of river water has already decreased considerably the rate of recession of the falls. It poses, however, a problem of great importance. The grandeur of Niagara Falls is due far more to the great volume of water plunging over the precipice than to the height of its fall. Fortunately, even those who want to use Niagara's power for the physical welfare of mankind are well aware of its unique interest as a wonder of nature. Under supervision and control by the two governments concerned and with modern knowledge of fluid dynamics and stream erosion, the scenic beauty of the Falls can be preserved, electricity can be produced within carefully specified limits, and the rate of recession can be reduced to a minimum. Niagara Falls will continue for countless generations to be one of nature's great gifts to man.

FALLS THAT STAY PUT

Very few of the world's great waterfalls resemble Niagara in their life history. Instead of maintaining their height and migrating upstream, most

falls are destined to remain approximately where they are and to decrease in height. The usual sequence is from a free-falling cataract to a series of cascades and then to rapids, as the vigorously eroding stream smooths out the irregularities of slope in its bed. The process can be seen in operation at many places.

The Falls of the Rhine near Schaffhausen, Switzerland, for example, are in reality a series of spectacular cascades although six or eight thousand years ago there was here an even more spectacular, free-falling cataract. The history of these falls, like that of Niagara, is intimately associated with events occurring during the glacial period, but there are notable differences. Far-spreading piedmont glaciers had descended from major valleys in the high Alps, and when they melted away from the Schaffhausen region, they left an undulating upland in which the preglacial valleys were filled with glacial drift. The Falls are at a place where the Rhine leaves the resistant rock of the upland to flow onward in a preglacial valley from which it has

Slender columns in Monument Valley, Arizona, are spectacular examples of the landforms occasionally produced by running water in an arid region. (David Muench)

117

quickly washed away much of the easily eroded glacial debris. The rocks at the base of the steep side-wall of the valley thus exhumed are as resistant as those at the brink of the falls; there is no undercutting of a precipice by backwash in a plunge pool, as at Niagara. Hence the Falls of the Rhine cannot recede upstream. Already the brink of the falls has been cut downward some eighty feet, and the steepness of the slope down which the water plunges has been notably lessened. Many thousand years will elapse before the cascades are reduced to relatively tame rapids, but that is their inevitable fate.

The Upper and Lower Falls of Yellowstone River in Yellowstone National Park, Wyoming, are similarly in the midst of this process whereby rivers smooth out the inequalities in their gradients. Here the river has eroded a typically youthful valley with sharp V-shaped cross section in a great mass of volcanic rocks, chiefly agglomerates and tuffs. For the most part these are not very resistant to erosion, but in their midst are more or less vertical bodies of harder rock resulting from the precipitation of mineral matter brought there by hot solutions ascending through cracks and fissures. Two such wall-like bodies of resistant rock happen to be athwart the course of the river in Yellowstone Canyon and are responsible for the two waterfalls. In the early 1950's the Upper Falls were 109 feet high, the Lower Falls 308 feet, but at each it is apparent that the brink has been lowered several feet by stream erosion in the last few centuries.

Victoria Falls of the Zambezi River in Rhodesia are due to a quite different geologic structure, but their fate will eventually be the same (Plate 36). The land mass there consists of a series of thick lava flows one above the other, forming an extensive plateau topped by volcanic mountains. Upstream from the cataract, the river is in a broad valley on the surface of the plateau. At the falls it plunges abruptly into a narrow gorge which it enters at the side rather than at the head. The gorge marks the position of a great fracture in the earth's crust along which the rocks were broken by crustal movements such as those which will be considered in Chapter Fourteen. It is not a single great rift but a zone in which the rocks were broken and therefore weakened by a multitude of displacements, with many closely spaced fractures running in all directions. The resistance of the shattered lava in this fractured zone contrasts strikingly with that of the rock on either side. The Zambezi, aided by a small tributary that enters the gorge at its head, has enlarged the fractures and removed the shattered rock to form the zigzag abyss below the cataract. Victoria Falls are more than twice as high as Niagara but the volume of water fluctuates greatly with the seasonal changes in the flow of the river. It is almost impossible to get a full view of these falls because of the narrowness of the gorge and the cloud of mist and spray that conceals its depths.

Most waterfalls in northwestern Europe, in the Alps, and in the western mountains of North America are a result of glacial erosion. It is more appropriate, therefore, to describe them in Chapter Nine than here. Several others among the world's most magnificent cataracts, like those just described, are due, however, to unequal resistance to river erosion.

Although Africa is far in the lead among the continents in the total amount of potential hydroelectricity that could be developed from its rivers, the South American Rio Paraña—called Rio de la Plata in its downstream course—has the most imposing array of giant falls. In the parts of Brazil and Paraguay drained by the Paraña and its tributaries, the bedrock is largely basaltic lava, extruded long ago to form a vast plateau, as flow after flow poured out from volcanic vents. At places the thick mass of tough rock has been fractured in long narrow zones by movements of this part of the earth's crust. The structure is similar to that at the Victoria Falls of the Zambezi. Wherever a river flowing across the upland encounters one of these zones of weakened rocks, it erodes rapidly downward and then plunges thunderously into the chasm. Guayra Falls, in the main channel of the Paraña on the border between Brazil and Paraguay, has more than a dozen cataracts averaging 110 feet in height along the sides of the main gorge. The Salto del Iguazu (Plate 35) is a series of cataracts in the Rio Iguazu, a tributary of the Paraña. The largest of these is about twelve miles upstream from the junction of the two rivers and is more than twice as high as the Guayra Falls but with a considerably smaller volume of water.

OLD RIVERS IN YOUNG CANYONS

The Grand Canyon of the Colorado River in Arizona is a definitely youthful feature of the landscape, a mile deep and only eight to twelve miles wide from rim to rim (Plate 30). But the Colorado has the dimensions of a river in old age. It averages three hundred feet in width and twelve feet in depth, and is one of the three longest rivers on the North American continent, the master stream in a drainage basin having an area of more than a quarter-million square miles. Why should such a river be in a youthful canyon? Should it not be flowing across a lowland plain, like the Mississippi or the Missouri?

To find the answer, we must note particularly the serpentine pattern of the stream course. The river and the canyon describe a great semicircle as they bend around Valhalla Plateau in the eastern part of Grand Canyon National Park, another similar though smaller curve around Havasupai Point, and then a loop around Great Thumb and Chikapanagi Points before leaving the Park at the west. It is the pattern of a trunk stream meandering across a lowland in the last stage of an erosion cycle. This is even more clearly shown by the Colorado and its tributaries upstream from the Grand Canyon. "The Goosenecks" (Plate 41) of the San Juan River near Mexican Hat, Utah, are probably the most spectacular example of a landform commonly displayed throughout the "Canyonlands" of Utah and Arizona. The San Juan empties into the Colorado in the Glen Canyon segment of that river in southern Utah. Its meandering course is typical of a stream in old age, but its steep-walled canyon is definitely a juvenile characteristic.

This is clearly a *two-cycle landscape*. One or two million years ago—some say seven million, but the smaller figures are more accurate—the Colorado and its major tributaries were flowing across a vast lowland. They had reduced much of their drainage area to an undulating plain fairly close to sea level. Like all elderly rivers they meandered in loops of varying dimensions across their floodplains. Then came a crustal movement, slowly but inexorably lifting the entire southwestern part of the continent, relative to sea level, by several thousand feet. The gradient of the lower part of the

Overleaf: Low mesas in the Badlands of South Dakota have edges like jigsaw-puzzle pieces due to differences in resistance to erosion of the horizontal beds of sedimentary rocks. (William A. Garnett)

119

Colorado was steepened, its velocity increased. It began to erode its channel downward toward the new base level of erosion, cutting a narrow gash in the plain across which it had formerly meandered. Vigorous erosion of its bed is characteristic of a youthful stream; in the vivid language of the geologist, the Colorado was rejuvenated.

The effects of rejuvenation appeared first near the mouth of the river, and gradually worked upstream as the youthful canyon was lengthened by erosion at its head. When the Colorado's canyon was thus extended past the mouth of a tributary stream, the latter was also rejuvenated and began to erode its own canyon which then was lengthened upstream by the same process. Wherever a river, such as the San Juan at "The Goosenecks," had been meandering in graceful, closely-spaced loops, the canyon was carved in nearly identical curves. The incised or entrenched meanders provide strong evidence for the theory of rejuvenation of those rivers and the two-cycle history of the region. "The Goosenecks" are spectacular because there has not been time enough, since rejuvenation worked its way upstream from the lower Colorado to that distant point, for weathering and slope wash to modify greatly the steep walls of the deeply cut meanders.

In comparison, the walls of the Grand Canyon have been magnificently sculptured by these agents of erosion with the help of numerous short, mostly intermittent, tributary streams fed by melting snow and occasional heavy rain draining into the canyon from either side. The Colorado itself cut only a narrow slot; the great width of the canyon is due to these other agents of geologic change. This, however, does not belittle the work accomplished by this mighty river. Upon it rests the responsibility for carrying away all the debris removed in the entire excavation of the gigantic chasm. On the average its swiftly flowing, turbulent, yellowish and reddish brown waters carry a half million tons of mud and sand each day beneath the Kaibab Bridge at the foot of Bright Angel Trail; probably a nearly equal weight of pebbles, cobblestones and boulders is swept or rolled along the river bottom at the same time.

The regional uplift of the land was locally exaggerated by a broad, gentle doming. The Kaibab Plateau beyond the north rim of Grand Canyon is higher above sea level than the Coconino Plateau beyond the south rim; both are higher than the upland surfaces on either side of the canyons traversed by the Colorado far upstream in Utah. The powerful river was able to cut downward at least as rapidly as the dome was arched upward and thus maintained its course while the canyon walls became higher and higher above it. For this reason the Grand Canyon is deeper than the Glen Canyon through which the Colorado flows in southern Utah and northern Arizona. Here also is an indication of the slow rate at which this kind of crustal movement may take place.

The sculptured walls of the canyon provide countless examples of the effect of varying resistance to erosion. The nearly horizontal beds of sedimentary rock are conspicuous, some in gentle slopes, some in cliffs and ledges, and some in "temples," "thrones" and buttes. They include, from the top downward, the light gray Kaibab limestone, forming the rim-rock cliffs and defending the tops of the temples against erosion; the buff Coconino sandstone, generally in steep slopes; the red Hermit shale, appearing in gentle slopes; the red Supai sandy shale and shaly limestone with its sculp-

122

tured buttes and variable slopes; the blue-gray Redwall limestone, almost everywhere stained red on its weathered surfaces and responsible for the "red wall" roughly midway between top and bottom of the canyon's sides; the buff Muav limestone in the steep slope just below the "red wall"; the greenish-gray Bright Angel shale, easily eroded to form the gentle slopes leading down to the Tonto platform; the brown, firmly consolidated Tapeats sandstone, protector of the rim of that broad shelf. Still farther down in the narrow, V-shaped inner gorge, is the exceedingly resistant metamorphic rock of the "basement complex," comprising the recrystallized sedimentary rocks of the Grand Canyon series and the more widespread Vishnu schist, shot through and through with stringers and dikelike sheets of pinkish granite.

Rare indeed are the places where one can riffle so many pages of the earth's diary as here. From the Tapeats sandstone up to the Kaibab limestone, many of the sedimentary beds contain fossils of marine animals or imprints of fernlike plants or footprints of primitive reptiles. Several of the Paleozoic chapters of the record are remarkably well preserved and have yielded a wealth of information about the history of this part of the continent during that time. The basement complex in the inner gorge is Precambrian in age and it too reveals a complicated sequence of events during that most ancient of eras. The Grand Canyon is not only a scenic marvel, attracting the artist and defying the descriptive powers of the writer; it is also a treasure-trove for the geologist, with important information still awaiting analysis.

A somewhat similar sequence, involving two or more cycles of erosion as a result of crustal movements, has been responsible for many spectacular features resulting from erosion by running water in other parts of the world. The Rhine Gorge in Germany, for example, displays all the characteristics of youth: V-shaped cross section; steep cliffs rising abruptly from either side of the stream; irregularities in the river bed, producing rapids separated by quiet reaches. (The rapids at the foot of the Lorelei Cliff, as well as those near Bingen, have been blasted away to improve navigation.) Above the cliffs and beyond the slopes with their terraced vineyards, an undulating upland plain represents the valley floor of the Rhine when it was a post-mature river in an earlier cycle of erosion. Crustal uplift rejuvenated the elderly river, terminated that erosion cycle, and stimulated the cutting of the winding gorge below the former valley floor in the present canyon-cutting cycle.

WHEN WATER RUNS THROUGH ARID LANDS

The startling contrast between landscapes where rainfall averages less than fifteen inches per year and those receiving at least twice that amount of precipitation is not merely a matter of different vegetation. The shapes of the hills, the contours of the valleys, the conspicuous erosion features all are distinctive.

Monument Valley in Utah and Arizona is a prime example of this (Plate 38). Its flat-topped, steep-sided mesas and buttes and its tall chimneys and pinnacles have been carved from horizontal beds of red sandstone and conglomerate. Some of the beds are unusually thick, others quite thin, but

In Cathedral Gorge State Park, Nevada, a mass of clay and silt deposited millions of years ago in a long-vanished Tertiary lake has been sculptured by rainwash into cathedral spires and robed figures. (Josef Muench)

the partings between the layers are planes of weakness to erosion. Similarly, the vertical joints are irregularly spaced but each makes the rock especially vulnerable wherever it occurs. The relations between these various planes and the shape and position of the "monuments" are clearly apparent almost everywhere. Differential weathering has contributed much to the sculpturing of the features of this unusual landscape, but erosion by running water is the principal agent in their origin. Desert rains commonly occur as torrential downpours lasting for only an hour or two. Flash floods rush down the normally dry washes and arroyos; short-lived streams carry an abundance of rock fragments and have great erosive power. The thinner beds of sandstone near the foot of each mesa or butte are less resistant than the thick beds overlying them; consequently the sides of a mesa continue to retain their steep, wall-like appearance even while its flat top dwindles in area. Thus each mesa tends to shrink in length and breadth until it is so small that it is called a butte. And buttes tend to become chimneys and pinnacles.

Vertical joint planes profoundly influence these progressive changes. The Mittens, for example, as viewed from the visitors' center maintained by the Navajo Tribal Parks Commission, really deserve their name. Each of the thumbs is separated from the rest of the butte by an open space, eroded away because of the unusually close spacing of joint planes in that part of the massive sandstone. Similarly, the chimneys and pinnacles are erosion remnants, originally surrounded by rock more closely jointed and therefore weaker than that composing them (Plate 39).

The Badlands (Plate 37) in South Dakota are quite different but they also are the result of erosion by running water in a comparatively dry region. Here the rocks are poorly consolidated, horizontally bedded sediments; the stratification is indicated by the outcrops of the thin sandy layers and the tawny gray, dark gray, and brown shaly beds contouring the intricately dissected, rounded hills and ridges. The occasional rains are of the cloud-burst type, and the agents of erosion are almost exclusively the flash floods in normally dry gulches, ravines and arroyos. A similar landscape dominates the Painted Desert in Arizona, where the poorly consolidated fine-grained sediments include many beds of brilliant red, green and yellow hues.

Rainbow Bridge (or Nonnezoshi Arch, as the Navajos call it), east of the Colorado River's Glen Canyon in southeasternmost Utah, is probably the longest and highest natural bridge in the world (Plate 40). It is not an opening through a steep-walled ridge separating distinct drainage basins and resulting primarily from differential weathering like the arches and natural bridges discussed in the preceding chapter. Instead it juts out from one wall of Bridge Canyon and terminates at the farther end in a gracefully curving buttress that rises abruptly from the canyon floor. Running water is mainly responsible for its sculpture. Bridge Creek is a two-cycle stream, like all the tributaries of the Colorado. Prior to its rejuvenation, which took place when Glen Canyon had been deepened by the Colorado's headward erosion, it meandered lazily across a plain, remnants of which are now the plateau surface and mesa tops on either side. As rejuvenation of Bridge Creek worked upstream, its meanders were entrenched like those of the San Juan at "The Goosenecks." Where the great arch is now, there was for a time an upland spur that formed the neck between two meander curves. Sweeping around those two curves, the flash floods of Bridge Creek under-cut the canyon wall on each side of that spur. Eventually the alcoves thus eroded were extended until they met, back to back. Bridge Creek no longer had to flow around the end of the spur; it now could straighten its course and flow beneath the arch it had sculptured.

Rainbow Bridge is by no means unique in its origin. Other natural bridges have been formed in the same way. Most commonly they are found where a fairly small stream, subject to flash floods, has been rejuvenated from a post-mature stage in the cycle of erosion to a canyon-cutting stage in a subsequent cycle. Conditions are most favorable where, as at Rainbow Bridge, meanders are deeply entrenched in flat-lying beds of sandstone of uniform texture and firm cementation. The agents of weathering are now the enemies that eventually will destroy the Bridge; but it is not weakened by close-spaced bedding planes or numerous joints and it will stand for many centuries to awe its beholders and challenge their comprehension.

7

Running Water: Nature's Builder

We have seen in the preceding chapter how the ceaseless sculpturing of the land by running water yields vast quantities of rock debris that is transported downstream—eventually to the sea. The load of yellow mud that gives the great Hwang-ho (Yellow River) of China its name adds an estimated two billion tons of alluvium each year to the delta that the river is building at its mouth. Similarly, the Mississippi River annually carries more than half a billion tons of mineral matter into the Gulf of Mexico. Nearly two-thirds of this travels suspended in its turbid water; about one-tenth is coarser particles that roll along the stream bed; the rest is mineral matter in solution. Yet the hundreds of billions of tons of sediment dumped annually by rivers into the oceans are probably almost equalled by the rock waste that streams deposit on land before they reach the sea. Some of these deposits build landforms of characteristic shape, and some of the landforms exert a profound influence upon the activities of mankind.

TORRENTIAL CONES AND ALLUVIAL FANS

Cone-like forms with symmetrical outlines and smooth slopes are conspicuous in almost every steep-walled, flat-floored valley in any rugged mountain range. The base of the cone lies on the main floor of the valley and the apex is commonly up in a gulch or ravine on the valley wall. Or the cone may be where a small tributary joins a larger river. The slope of the cone may be steep or gentle; it may nestle close to the valley wall or spread far out across the valley floor. Its location and shape and the type of material in it provide clues to its origin.

Various factors cause streams to deposit part or all of their load of sediment at certain places. Thus, a reduction in velocity or volume of water decreases transporting power, as does also a change in shape of a stream channel that increases friction between the moving water and its bed. All three of these factors help to produce the conical landforms described above. The gradient of the stream bed changes abruptly from steep to gentle at the break between valley wall and valley floor; the stream's velocity drops suddenly at that point. At the same place, the narrow, deep channel cut by swiftly flowing water in the gulch or ravine becomes wider and shallower; increased friction further retards the stream. As soon as a considerable deposit of rock fragments has accumulated, some of the water will flow into the

open spaces between the fragments; the volume of the surface flow is thus reduced. The symmetrical shape of the cone is due to frequent shifts of the stream course below its apex; some of the water will run down the slope in one direction and some in another as the stream clogs its bed and separates into several channels. Thus the deposits are evenly distributed and the symmetrical form is maintained while the cone grows.

Steeply sloping, conical accumulations of stream deposits are generally composed of coarse debris, often including boulders and angular blocks a foot or more in diameter, as well as cobblestones and gravel. Such materials could have been transported only by a torrent from a sudden downpour or the rapid melting of snow during a warm day in spring. Perhaps too, an avalanche of snow was funneled down the mountainside, hit the apex of the cone and spread debris upon its sides. The name *torrential cone* is applied to such landforms. Transitions between them and the talus cones mentioned in Chapter Four are sometimes seen.

Extending far out across Death Valley, California, are many sloping alluvial fans, like this one at the mouth of Hanapah Canyon. (Spence Air Photo)

127

Many torrential cones in Alpine valleys, like the one at right, are formed mainly by meltwater from glaciers high on the mountain.
(Photo Klopfenstein)

The conical landforms built by fairly large streams with more uniform volume often spread far out across the main valley floor in a comparatively gentle slope. Moreover, they consist largely of silt, sand and gravel, with only an occasional large fragment of rock. For them, the name *alluvial fan* is widely used. There is, of course, every gradation from torrential cones to alluvial fans. Many factors make alluvial fans especially attractive for villages and cultivated fields. Both natural and artificial drainage are efficient and easy on the gentle, even slopes. Their height above the floodplain, though moderate, generally suffices to give protection from floods; and the silty alluvium ordinarily provides fertile soil. Many of the villages, towns and small cities in the larger valleys of the Alps are situated on alluvial fans; Visp in Switzerland and Chamonix in France are good examples.

PIEDMONT ALLUVIAL PLAINS

In many parts of the world, mountain ranges that rise abruptly from lowland plains or broad, flat-floored basins catch most of the rain and snow, and little or none falls on the lowlands. Each of the many youthful streams, busily engaged in carving valleys in such mountains, builds an alluvial fan at the mouth of its gulch or ravine and neighboring fans coalesce to form a *compound alluvial fan*. If this landform is large, it is often referred to as

128

a *piedmont* (literally, foot of the mountain) *alluvial plain.* Where the streams leave the mountains and start across the lowland, their volume drops significantly because of evaporation in the arid or semiarid climate there and because of absorption of water by the thirsty ground of the fans. Moreover, runoff in such mountains generally fluctuates greatly due to dry and wet seasons. Many of the streams flow intermittently and even streams that are permanent in the mountains may disappear before they get very far out on the piedmont plain. Such variations in the volume of running water are probably as responsible for the stream deposits as is the change in gradient of the stream beds.

Piedmont alluvial plains are conspicuous features of the landscape on every continent. They stretch eastward from the front of the Andes Mountains in eastern Bolivia and northwestern Argentina; they border the south face of the Atlas Mountains in Algeria and Morocco. Much of Los Angeles, California, occupies one of them, veneered with alluvial deposits carried by rivers from the San Gabriel Mountains, north and northeast of the city. The extensive alluvial plains in the fertile San Joaquin Valley in California, built outward from the encircling Temblor Range, Tehachapi Mountains, and Sierra Nevada, are now under widespread irrigation, as are other such deposits in similarly dry regions. These valleys are especially amenable to cultivation because gravity alone conducts water from the mountain canyons and spreads it over all parts of the smooth, gently sloping surfaces. In more humid climates, irrigation may not be necessary. There the thick, porous alluvial silt serves as a reservoir for ground water and thus contributes to the fertility. Large parts of the fertile valley of the Po River in northern Italy, for example, are really piedmont alluvial plains built by streams flowing south from the Alps.

FLOODPLAINS

Late in the mature stage of the erosion cycle, flat bottomlands usually develop in the valleys of all rivers. These are commonly veneered with sediment deposited by the stream. Much of this comes from flood waters and this portion of the valley is often referred to as the floodplain. The river makes these deposits on its floodplain because it was "overloaded" in its upper reaches with more sediment than it could transport on the gentler gradient in its lower course. Even slight changes in gradient along the stream bed upset the delicate balance between transporting power and velocity. Streams flowing in broad, shallow channels often deposit sand bars that appear as islands in times of low water. If a stream is intricately subdivided by many sand bars and islands, constantly changing in pattern as the water shifts from a clogged passageway to a more open one, it is said to be "braided." Once deposition of sand bars begins, the bars themselves tend to accelerate the process; the many distributary channels of a braided stream increase the friction that retards the flow—the same amount of water flowing over the same gradient in a single deep channel could carry a much greater load.

Where streams form floodplains of fine silt, the channels are generally very sinuous or meandering (Plate 44) rather than braided, although braided streams may also meander. Many loops or oxbows of abandoned channels mark the floodplain on either side of the main channel. These may be filled in and dry, or swampy and clogged with vegetation, or they may take the form of oxbow lakes. The highest land is ordinarily adjacent to the stream channel; farther away on either side the surface of the floodplain is generally somewhat lower. These higher strips are *natural levees* built by deposits made during floods. As the floodwater leaves the swift current of the deep channel to spread thinly across the plain, its velocity drops and it deposits much of its sediment. The slightly higher land of the natural levees is commonly more valuable for agriculture and town sites than the lower, often swampy, surface of the floodplain farther out. In spite of recurrent floods the nearly level, easily tilled surfaces of silty floodplains, with their fertile alluvial soils, are among the most highly prized agricultural lands of every continent. The floodplain of the Nile, for example, is only about ten miles wide throughout the seven hundred miles of its length from the cataracts at Aswân downstream to the Nile delta, yet it and the delta support nearly all of Egypt's dense population.

The much wider floodplain of the Mississippi measures 170 miles across near Memphis, Tennessee. Its alluvial deposits, like those of other large rivers emptying into the sea, are two to four hundred feet thick and extend downward far below sea level. At most places, the finer silt extends from the surface to depths of only a few score of feet and the lower layers of sediment consist of coarse material with many sand and gravel beds. Past geologic events explain this structure. During each of the four episodes of continental glaciation in the Great Ice Age, sea level throughout the world dropped three or four hundred feet as a result of the temporary imprisonment of water in the millions of cubic miles of ice. This lowering of sea level rejuvenated the Mississippi and it eroded its bed downward. Having a steeper gradient, it could transport sand and gravel as well as silt. Four times the ice sheets melted and sea level rose to its present relative position. Each time the velocity of the river promptly decreased. It first dropped its load of gravel and sand, and then deposited sufficient silt to raise its floodplain to the level of the sea at the river mouth. In other words, the broad, nearly level bottomlands are not the result of simple downward and lateral erosion by a river in old age. Instead, this floodplain is the result of four separate epochs of deposition by running water. Each time the river filled with sediment the lower part of a deeper valley which it had earlier eroded.

DELTAS

Floodplain deposits of a river that empties into a sea or lake merge downstream with the sediments of the delta at the river mouth—if there is a delta there. The term "delta" was first applied by Herodotus in the fifth century B. C. to the alluvial plain through which the branching channels of the Nile River debouched into the Mediterranean Sea. On his map that region had the triangular shape of the Greek letter, delta. Geologists now use the term to include the entire deposit formed either above or below water level at or near the mouth of a river. Few such deposits really have the shape of the Greek letter.

The "standing" water of a lake or sea quickly checks the forward movement of a river which promptly drops much of its load of silty or sandy sediment at its mouth. Clay particles and colloids (particles so small that they cannot be seen with an optical microscope) often remain in suspension in lake water for a long time and therefore may drift far before they come to rest. But the salts in sea water cause them to flocculate—that is, to collect in small masses or lumps, which quickly settle to the bottom. Consequently, marine deltas ordinarily contain a higher percentage of such ultrafine particles of mud than lake deltas. Only where the river-borne sediment is widely dispersed by strong currents or tidal flow will a river fail to build a delta at its mouth.

The Matanuska River deposits much of its sand, silt, and clay along the sides of its channels as it flows into the tidal estuary at Cook Inlet, Alaska.
(Steve McCutcheon)

131

The shape of the deposit is determined by many factors, such as the slope of the sea or lake bottom, the pattern of the shoreline, the direction and velocity of currents in the body of "standing" water, the velocity and volume of the inflowing water, and the quantity and character of the arriving sediment. Each delta has unique features, even as each person in a group is different from all the others. No other delta, for example, has the "bird-foot" pattern of the outermost tip of the Mississippi Delta. It is in striking contrast to the relatively smooth bulge of the Nile Delta along the Mediterranean shore or the even less conspicuous extension of the Po Delta into the Adriatic.

The mechanics of delta construction are generally apparent wherever a fairly swift stream empties into a lake or a long narrow bay, sheltered from the open sea, as in the lakes and fjords of Norway (Plate 42). No line of demarcation can be drawn between the downstream end of the floodplain and the part of the delta that is above water. Beyond the shoreline, the top of the delta extends outward in shallow water, but somewhere, near or far from shore, its outer margin is generally marked by an abrupt deepening of the water. Here partially submerged fingers of sand or silt may reach out into the deeper water; they mark the roughly parallel sides of the jet stream of river water before it mingles completely with the quiet water into which it is flowing.

This frequently observed pattern of delta fronts is shown especially well by the so-called passes at the outer margin of the exposed portion of the Mississippi Delta. The "passes" are the many channels through which the Mississippi distributes its water on the way to the Gulf of Mexico. They change frequently during times of flood, when the river breaks the natural levees it has built and abandons an old course for a new one. At one time, navigation from the Gulf into the river, tricky at best, was further complicated by the unpredictable deposits of silty bars in the entrance to a "pass." This problem was tackled by an American engineer, James B. Eads, a hundred years ago. He solved it by constructing parallel jetties along the entrance to the "passes" so that the narrowed stream scoured out its own channel and carried the sediment out to sea. It is evidently far better to humor Nature than to fight her. All that was subsequently required was an occasional extension of the Eads jetties to make sure that the sediment deposited beyond their ends was laid down in water so deep that it would not interfere with navigation.

More difficult and more ambitious has been the task of the people of the Netherlands as they have sought for centuries to make their delta a prosperous living place. Nearly all of that "Low Country" is on the coalesced deltas of the Rhine and Maas Rivers, each of which divides into many distributaries that carry its waters into the North Sea. The flat surface of the deltas, just above sea level, must be drained; stream channels must be stabilized and kept navigable; portions of the delta surface, just below sea level, must be cut off from the sea by dikes and pumped free of water; an intricate network of canals must be constantly regulated for drainage and transportation. In the construction of many of the polders—the local name for lowlands diked and drained to become highly productive agricultural land—the Dutch have taken advantage of the long sandy beach. Backed by sand dunes, it stretches in a sweeping curve northeastward from the Hook of Holland past The Hague to Den Helder and then is represented by the discontinuous line of the Frisian

Islands as it curves eastward. It was built offshore along the seaward margin of the delta by the strong currents and powerful waves of the North Sea and serves as a natural dike to facilitate the salvaging of land between it and the inner part of the delta above sea level. Even with this assistance from nature, the battle against sea and river is arduous and continuous. The Dutch are winning their struggle because of their study of the behavior of running water and the processes of shoreline change, and because of the wisdom of their engineers in applying many different kinds of knowledge to specific operations within a well-designed program of reclamation and conservation.

The growth of some deltas is surprisingly rapid. The delta of the Rhone at the opposite end of Lac Leman from Geneva, Switzerland, is a well-known example. The Rhone is heavily laden with silt and sand supplied by meltwater from many glaciers in the Alps. As a consequence, a delta has formed in the eastern end of the lake—sometimes incorrectly referred to as Lake Geneva—which extends more than twelve miles into the original lake basin.

Shrinkage cracks appear on the parched surface of silt and mud deposited by the Colorado River at the head of Lake Mead in Arizona. Seen here are two depressions resulting from gas bubbles that escaped while the mud was still water-soaked. (Tad Nichols: Western Ways Features)

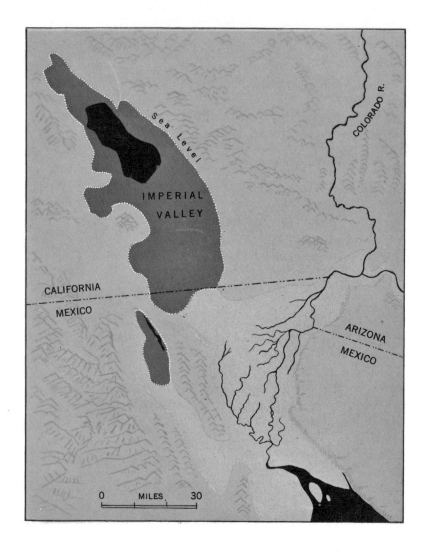

A delta (intermediate gray) formed by the Colorado River in California, Arizona and Mexico has built up a dam across the Gulf of California (shown in black in the lower right corner). The water in the Gulf north of the delta has evaporated and left the Imperial Valley below sea level, with the Salton Sea occupying its lowest part.

One mile of this has been built since Roman times. The fertile soil of the deltaic alluvium is extremely valuable for agriculture, and the scenic beauty of the lake region is unaffected as yet by the deposits in this narrow end of the lake.

In contrast, the rapid construction by the Colorado River of a delta in the upstream end of Lake Mead is a matter of some concern. Lake Mead is an artificial lake, its waters impounded by Hoover Dam. Since the completion of the dam in 1936, the Colorado has partly filled the narrow gorge at the upper end of the lake with fifty miles of deltaic deposits. More significantly, the cold, muddy water of the river sinks to the bottom of the somewhat warmer lake and moves far beyond the end of the delta to deposit its load of fine clay particles on the lake floor all the way to the dam. These deposits have already reached a thickness of 135 feet against the dam even though the river empties into the head of the reservoir 75 miles away. So great is the capacity of this reservoir, however, that it will still be useful a hundred years from now, despite the potent efforts of running water to fill it with stream-borne sediment.

Long before men began to interfere with the Colorado, it had built a huge delta at its mouth in the Gulf of California. The river originally entered this long, narrow gulf near the site of the present city of Yuma, Arizona, some 150 miles south of what was then the northern end of the Gulf. The Colorado Delta, topped by a vast alluvial fan, extended clear across the Gulf, completely cutting off its northern part from access to the sea. The river stabilized its course to the south, presumably during the last episode of lowered sea level in the Great Ice Age. The lake north of the deltaic dam evaporated in the dry climate of southeastern California until all that was left was the saline water of the Salton Sea with its surface about 250 feet below sea level in the lowest part of the Imperial Valley. The floor of the valley slopes gently downward from sea level near the California-Mexico line to the shore of the Salton Sea, forty miles to the north. Its fine alluvial soil needs only water to make it fertile, and the Colorado River is supplying the water through a network of irrigation canals so that the Imperial Valley now contains hundreds of square miles of cultivated lands.

The situation in the Gulf of California is not duplicated anywhere else, so far as I know, but the separation of a body of water into two parts by the construction of a delta is not rare. Two lakes, Buttermere and Crummock Water, in the Lake District of England, occupy parts of the floor of a deep valley blocked by glacial deposits at its lower end. Formerly a single lake, they are now separated into two parts by a delta topped by an alluvial fan deposited by a stream flowing in from the eastern side of the valley.

Similarly the Lake of Thun and Lake Brienz in Switzerland were formerly a single body of water. It was bisected by the delta of a stream fed by debris-laden water from the melting glaciers in the nearby Alps. The well-known, appropriately named resort town of Interlaken stands on this deposit of river-borne silt, sand and gravel.

8

Where the Land Meets the Sea

Almost everyone who has lived near the sea is impressed by the way the shore is constantly changing. In few other places do such powerful forces alter the landscape from day to day. Variety as well as beauty characterize the scenery of most seashores: waves ceaselessly pounding craggy headlands; booming surf breaking on ever-shifting beaches and offshore bars; silent marshes that are half land and half water. It is no wonder that both in song and story the seashore has had such an important place in the literature of many peoples. Nor is it surprising that we are fascinated by explanations of the diversity of beaches and crags, offshore islands and sheltered coves, sandy bars and tidal flats. Such explanations not only quicken the imagination but also have great practical significance, as in the protection of coastal property and the improvement of harbors.

WAVE EROSION

Waves and currents exercise immense power as agents of geologic change. The breaking waves that dash high against a rocky cliff are not only awe-inspiring; their efficiency in eroding the land calls for serious study. Wave erosion involves several factors. The height of waves from crest to trough, their length (the distance between successive crests), and their period (the interval between arrival of successive crests) are determined by the velocity and duration of the winds that generate them and by the fetch (the area of water across which the winds blow in a constant direction). As waves travel toward shore from deep water, their physical characteristics change. When they reach a depth equal to half the wavelength, they "feel bottom" and enter a transitional zone; the wave velocity and length begin to decrease, the wave steepens on its shoreward face and grows higher. In the breaker zone these changes are accelerated so that the waves may become several times as high as the deep-water waves. Continuing into shallow water where the depth is only a small fraction of the wavelength, the breaking and broken waves hurtle forward against all obstructions with an impact like that of a battering ram. The onslaught of the water often traps air in cracks and crevices in the rock; that air may be alternately compressed and released with explosive violence. Like streams of running water, the waves and currents in seas and lakes work most efficiently when provided with adequate tools. These are readily at hand in the turbulent water of broken waves—the sand, gravel,

shingle, and larger rock fragments present along every strand. Hurled by waves or swept by currents against exposed rock faces, they grind away the solid rock and are largely responsible for the sculpturing of most wave-cut cliffs.

Although waves and currents are similar to rivers in their dependence upon the tools of erosion, the results of their work differ notably from that of rivers. Erosion by waves and currents resembles the work of a gigantic buzz-saw whirling horizontally, approximately at the level of sea or lake. No matter how high the wave-cut cliff, the direct action of waves and currents is confined to the narrow horizontal zone between the highest and the lowest level of the sea or lake. Where tides are great and occasionally reinforced by strong onshore winds, this zone may be twenty or thirty feet high, at best only a trivial fraction of the height of most land masses undergoing erosion.

Waves attacking a smoothly sloping coast first produce a "nip," a low cliff with a narrow wave-cut platform, in front of which, on the sloping floor of the sea, is a thin veneer of sediment, the debris from the wave-cut nip, washed outward by the undertow. As the attack continues, the cliff retreats and the wave-cut platform extends landward. If the rocks offer uniform resistance to erosion, the cliff will tend to be a smooth wall, like the Chalk Cliffs of Dover which stretch in an almost unbroken line for miles along the English Channel. If, however, the rocks vary in strength from place to place,

On the California coast between Monterey and Carmel, onrushing waves as well as backwash have rounded angular blocks into boulders.
(David Muench)

137

the stronger ones will stand out as headlands, and the weaker will retreat more rapidly to produce fissures, clefts and coves. Since this is the more usual condition, wave-cut cliffs generally display an intricate pattern of precipitous promontories separated by coves and chasms, such as those along the Monterey Peninsula in California or the shore of the Cape of Good Hope, South Africa (Plate 51).

In general, the same characteristics determine the resistance of rocks to erosion by waves and currents as those that influence the rate of erosion by running water. The spacing and angle of joint planes and bedding planes, the hardness and chemical composition of the minerals, and the degree of consolidation are the important factors. Some combination of these explains the picturesque features of every wave-cut cliff. Thus, *sea caves* (Plate 54) are hollowed out by storm waves at the foot of many shore cliffs at places where the rock is less resistant than elsewhere. These may become veritable tunnels, or a part of the roof may collapse or be worn away to leave a *sea arch*.

Along coastlines that were originally irregular or have been made irregular by wave erosion, the attack is concentrated where incoming waves converge. Thus a small cleft in a cliff tends to grow larger and to extend more deeply into the land as the surging water funnels into it. The numerous "thunder holes" and "spouting horns" along many shore cliffs bear noisy testimony to the power of converging waves. For the same reason, a narrow steep-sided arm of rock may be attacked at the side more effectively than at its outer end. Thus the water may carve arches along the arm, as it has at Santa Cruz and San Diego, California (Plates 55 and 52).

At many places along rugged coasts, pinnacles of erosion-resistant rock project above the water just offshore, such as those at Pte. du Van on the coast of Finisterre in northwest France, or Cannon Beach, Oregon (Plates 53 and 45). Formerly parts of the mainland, they have been isolated by wave erosion and now rise from the water-covered, wave-cut platform. Such islands, especially if they are steep-sided and chimney-like or slender towers or pinnacles, are called *stacks*. Not all steep-sided, offshore islands, however, were formed in this way. Many islands along irregular shores are the former hills of a land drowned beneath the waters of a sea that advanced across it. Even if later wave erosion steepened their sides, they are not truly stacks since they were not separated from the mainland by waves and currents during the retreat of a wave-cut cliff. The famous Percé Rock, along the side of the Gaspé Peninsula in Quebec, for example, was formerly a hill rising above the lowland. The water of Chaleur Bay surrounded it when sea level rose everywhere at the close of the Great Ice Age. Since then, wave erosion has decreased its area and made its sides precipitous. The waves also "pierced" it with the sea tunnel that gives the rock its name. But erosion did not isolate it from the mainland, for no wave-cut platform extends from its base to the Gaspé shore.

The sea rose to its present worldwide level when the continental ice sheets finally melted away, about six thousand years ago. Except for local changes following movements within the crust, sea level has fluctuated only slightly since then. Nearly all of the changes wrought by waves and currents along ocean shores around the world have been made during this short period. The extent of these changes demonstrates impressively the potency of those

agents of erosion. Wherever the land rises steeply from the sea, there is almost always a wave-cut cliff that permits an approximate estimate of the extent to which the land has been worn away. In the archipelago of the Cyclades in the Aegean Sea, for example, each island displays a wave-cut cliff from a few feet to a hundred or more feet in height. Projection of the slope above the cliff out into the sea provides the measure of the erosion at that place.

The rocks of the Cyclades are sedimentary and not very resistant to erosion but those islands are sheltered in a rather quiet sea. In contrast, the coast of Finisterre is exposed to the full force of Atlantic gales and is pounded by powerful waves. Here, near Penmarch, low tide exposes a wave-cut platform nearly half a mile wide, cut in resistant granite. Scattered over its surface are occasional boulders and pebbles left by the receding tide and temporarily stranded while in transit from the foot of the wave-cut cliff to deeper water offshore. There are deposits of sand and mud in hollows and patches of seaweed in many places. The pools of water in basin-like depressions shelter crabs, mussels, sea urchins and other denizens of the near-shore shallows; the people in the area collect a considerable part of their food supplies here.

The gleaming white chalk cliffs near Étretat on the French shore of the English Channel are slowly retreating before the onslaught of waves and longshore currents. (Ray Delvert)

CHANGES OF SEA LEVEL

Movements within the earth's crust have elevated the land at some points so that features sculptured by waves and currents now stand scores or hun-

Most coral atolls like Mille, one of the Marshall Islands in the Pacific Ocean (as seen from the air), are irregularly shaped rather than neatly circular. (U. S. Navy)

dreds of feet above sea level. Changes of level with respect to the surface of the sea result from the relation between two factors: local movements of the earth's crust, upward or downward, and worldwide fall or rise of the sea surface as a result of expansion or contraction of continental ice sheets and glaciers. Wherever a remnant of a wave-cut platform stands well above the reach of storm waves at high tides, the uplift of land due to crustal movement greatly exceeded the rise of sea level as the last ice sheets dwindled.

Such elevated platforms can be seen at several places along the Pacific shore of the two Americas (Plate 50). Good examples are found on the coast of Oregon in the United States, and of Baja California in Mexico. Even more extensive are the so-called *tablazos* in northern Peru, standing steplike above the shoreline at the foot of the western Andes. Each is a wave-cut platform, beveling the tilted and folded beds of sandstone and other sedimentary rock, and thinly veneered with alluvial deposits and wind-blown sand. Each terminates seaward at the top of a wave-cut cliff, formed when the next-lower tablazo was being cut. The lowest step is generally a few score feet above the modern strand; the highest stands more than three thousand feet above sea level in some places. In this almost completely arid region, the record of wave erosion that took place a hundred thousand and more years ago has not been destroyed by running water as it would have been in a humid region. Thus the several successive uplifts of the earth's crust took place much farther back in geologic history than might be inferred from the apparent "freshness" of these elevated remnants of the former sea floor.

There are other raised platforms farther south along the coast of Chile, but none so spectacular as the Peruvian tablazos. Likewise, small remnants of

uplifted marine platforms may be seen at a few places along the shores of other continents. Their absence does not mean that relative uplift of the land has not occurred in recent geologic time; it may be due to the complete destruction of the record of uplift by subsequent erosion.

SHORE DEPOSITS

Waves and currents are not always agents of destruction; like rivers, they do constructive work in many places. Indeed, the beaches they build are far more widespread than wave-cut cliffs and shore platforms.

Beaches the world around have certain common characteristics that permit the use of a general terminology. The *shore* is technically the zone between the lowest water line and the line marking the farthest reach of attacking waves. The region seaward is known as the *offshore;* on the landward side is the *coast.* The beach ordinarily has both a *backshore* and a *foreshore,* the former higher than the latter and reached by waves only during storms or when tides are high. There is commonly a marked change in slope between the backshore and the foreshore which extends to the low water mark and is the portion of the beach traversed by the uprush and backwash of ordinary waves. It often displays a steeper seaward-dipping *face* merging downward into a more gently sloping *terrace* over which the waves break and surge. Along lowland coasts the backshore usually slopes landward so that there is a distinct *beach ridge,* low and rounded, which marks the upper limit of effective action by storm waves when tides are highest and water is

Longshore currents in Icy Bay, Alaska, have shaped the beach (near top of photograph) and its continuation as a bay-mouth bar. They have also curved the spit (lower left) into a hook.
(Bradford Washburn)

141

Lines of blackened seaweed on a beach at Partridge Island, Nova Scotia, record the progressively lower reach of high tides, from the spring tide when the tide is highest and the moon is new or full, to the neap when the moon is in first or third quarter. (Benjamin M. Shaub)

piled against the shore by strong winds. Beach ridges, backshores and adjacent coastal lands are often covered by dune sand moved by wind from the beach face when it is dry. Offshore, *longshore bars* and *longshore troughs* commonly run parallel to the beach. Longshore bars generally form on the shallow bottom beneath the line of breakers where incoming waves plunge forward to surge across the foreshore terrace. These bars may be exposed to view at low tide.

Because continuous, easily observable geologic activity takes place on the foreshore, it is of great interest to every lover of nature. Here the rhythmic movement of waves and backwash supplies the energy for a grinding operation like that of a mill. Angular fragments of rock, rived from cliffs and crags, are rounded into cobblestones and shingle, cobbles are reduced to pebbles, and pebbles are ground down to sand grains. The water sorts smaller particles from coarser and drops each in zones or bands that often shift as wave dimensions and velocities change from hour to hour or season to season.

Along most coasts the character of waves and the direction from which they come vary with the season. This may result in a notable difference in shore processes. Except in the tropics, the less violent waves of summer generally build beaches seaward. The more vigorous waves of winter storms cut back the beaches. There are also irregular shifts between cut and fill as a result of stormy conditions that may occur at any time and in any relation to the tide schedule. Thus a beach is not a stable deposit and can fluctuate rapidly in width and height.

Backwash from waves tends to transport particles directly outward to deeper water, but much of the material in transit is also moved along the shore (more or less parallel to it) by longshore currents. Such currents are generated by waves that approach the shore obliquely and are deflected when they "drag bottom." Their direction and velocity are influenced by the contour of the shoreline and the undulations of the sea and lake floor. Their effects are apparent in any careful inspection of the shape and position of the beaches (Plates 47, 48 and 56).

142

Gently curving *crescent beaches* frequently stretch from headland to headland along irregular shores. If the projecting points of land are close together, such beaches will be short; they are then called *pocket beaches* because they are tucked away within a "pocket" between cliffs. Where the floor of lake or sea shelves gently outward toward deeper water, *barrier beaches* are built offshore, their location determined by the place where the progress of the larger incoming waves was impeded by the shallowing water. A beach extending from land and terminating in open water is a *spit;* often its outer end will curve to form a *hook.* The Hook of Holland is the classic example but Sandy Hook, on the New Jersey shore at the entrance to New York harbor, is much larger; another curves from the tip of Cape Cod near Provincetown, Massachusetts, into Cape Cod Bay. A spit that extends part way across the entrance to a bay is a *baymouth bar,* whereas a beach near the head of a fairly long, deep bay is a *bayhead beach.* A beach connecting an island with the mainland or with another island is a *tombolo;* Monte Argentario, on the west coast of Italy, is "tied" to the mainland by two tombolos.

DEFENDING THE LAND AGAINST THE SEA

From time immemorial, men have tried to protect valuable coastal land from erosion by waves and currents. Construction of seawalls and breakwaters is costly and many have proved inadequate to withstand the battering of the unusually powerful waves of violent storms. It has generally been more effective to dump huge blocks of granite in front of the seawall or on the foreshore of a beach. But more accurate knowledge of shore processes has become available in recent years and engineering techniques have been developed to take advantage of the constructive aspects of shore processes rather than merely to oppose their destructive action. In most places the construction of groins or jetties provides the best protection. These are built at right angles to the shore and extend from the foreshore face clear across the foreshore terrace. Their location and spacing are selected after detailed investigation of the characteristics and behavior of the longshore currents at the particular locality. Beach erosion is now a special study of engineers and government bureaus in many parts of the world.

CORAL REEFS

Nature has ways of protecting coastal lands. Coral reefs fringe many islands and parts of continental borders in and near the tropics. The corals and their allies, especially the coralline algae, are quick to repair damage between storms and thus maintain their defenses against the waves. Coral polyps are tiny animals, far down on the scale of living organisms. Each resides in a cup constructed by its epidermal or outer cells from the calcium carbonate dissolved in the surrounding water. The aggregate mass of the cups of countless coral polyps in a colony makes the rocklike substance known as coral; the numerous species producing it account for the great variety of form and color. Associated with the typical reef-building corals are still lowlier one-celled plants, the so-called red algae, which secrete calcium carbonate within

143

and between their cell walls. The vast number of these little animals and plants, crowded together in favorable environments, form what are known as coral reefs. They also contribute notably to the making of limestone.

Corals are exclusively marine. Most persons associate them with tropical waters and this is understandable, but some varieties can live in cold water and even at great depths. There are coral banks and patches in the North Atlantic Ocean as far north as the coast of Norway. The typical reef-building corals are restricted to water warmer than 22 degrees Centigrade (71.6 degrees Fahrenheit) and to depths not greater than twenty-five fathoms (150 feet). Beyond that depth the light is too weak for the photosynthesis essential to the microscopic algae (known as zooxanthellae) with which corals live interdependently. True coral reefs are found only within 30° of the equator. They are most common in the Pacific and Indian Oceans but also occur in the Atlantic Ocean and the Red Sea. Very few are present on the western borders of continents, even in the tropics, because of the upwelling of cold water from the depths along these coasts. The coral reefs found in ancient limestones as near the poles as Greenland suggest climates and oceanic circulation quite different from those of today. This idea, however, should be accepted with caution. The coral reefs in Paleozoic rocks were made by creatures quite different from those responsible for Mesozoic and Cenozoic reefs. Nor can we be sure that Mesozoic and early Cenozoic reef-builders were limited precisely to the restricted environments of their more recent descendants.

Coral reefs fall into three classifications: *fringing reef, barrier reef, and atoll*. The fringing reef occupies the foreshore and terrace along the shore. The backshore is usually formed of coral fragments, ripped from the reef by storm waves, rounded into coral sand, sorted by size and left by the uprush of the almost spent breakers. Barrier reefs are separated from the coast by lagoons of various widths and depths; these depths are invariably less than that of the water a short distance out on the seaward side of the reef. Atolls are irregularly circular coral reefs surrounding a lagoon in which no land projects above the water. They vary in diameter from less than a mile to about twenty miles. In all atolls, the ring of long, narrow, low islands surrounding the lagoon is broken by channels through which the tides ebb and flow. Some of these are as deep as the lagoon and provide access to its shelter in time of storm. The depth of large lagoons rarely exceeds fifteen or twenty fathoms and generally the crests of underwater, coral-covered mounds or knolls scattered irregularly across the lagoon floor rise well toward low-tide level. The exposed portions of the atoll consist of reef fragments, more or less rounded and in many places cemented together to form a fairly firm rock. Dunes of wind-blown coral sand may likewise be firmly consolidated or they may be loose and on the move.

In the western Pacific, hundreds of atolls rise abruptly from the ocean floor two or three thousand fathoms deep. This poses the problem of their origin; the first of the corals that built them could not possibly live in water so deep. Several years before Charles Darwin set forth his theory of evolution he proposed a theory of the origin of atolls to solve the problem. According to this theory, fringing reefs are first constructed approximately at sea level around a volcanic cone built upward from the sea floor so that its upper part is an island. The sea floor and volcanic cone then slowly subside or

45. Inequalities in resistance to erosion explain the many "stacks" rising from shallowly-submerged Cannon Beach, Oregon. The swirling backwash from each spent wave rounds the rock fragments riven from the cliffs. (Robert Clemenz)

144

46. This inlet, on the coast (above) of Baja California in Mexico, has a pocket beach at its head; beyond is a slender spire or stack, isolated by wave erosion. (John Hodgkin: Western Ways Features)

47. Along a shoreline of emergence, as illustrated by the one near Monte Leon (right), Argentina, an elevated platform bordered by a wave-cut cliff often indicates the former position of the sea. (Hal Cerni)

48. Shorelines of submergence, produced by a sea that invades a stream-carved landscape, are characterized by bays and coves between promontories and headlands, as on the Cote d'Azur (facing page) in France. (Gerhard Klammet)

49. Burnished by a low-riding sun, the mud flats exposed by the ebbing tide of Bristol Bay (left), Alaska, display the rills and ripple marks often seen on bedding planes of shale. The man is standing beside a dead Beluga whale stranded by the swift ebb of the tide. (Steve McCutcheon)

50. The California coast near San Simeon (right) is bordered by a shoreline of emergence, as shown by the elevated platform in the distance. Erosion of the wave-cut cliff has been irregular because of the varying resistance of the rocks. (Mary S. Shaub)

51. Although the sea is calm, the surf relentlessly sculptures the foot of the cliffs at the tip of the Cape of Good Hope (below) in South Africa. (Laurence Lowry: Rapho Guillumette)

52. Low tide at San Diego, California, exposes a wave-cut platform on which stands a tower-like structure known as a perforated stack (above); in time the waves will undermine the tower until it collapses. (Paul Miller)

53. Similar wave erosion, in the zone between high and low water, has shaped the cliffs on either side of Pt. du Van (left), on the coast of Brittany, and left its most resistant rocks standing in the form of stacks. (Kirtley F. Mather)

54. Sea caves and sea arches are common features of shores where rocks of varying durability are being eroded. The Blue Grotto in the island of Malta (above) can be entered when wind and water are calm; its abundance of underwater plants and plantlike animals is amazing. (Josef Muench)

55. The arches in the promontory at Arches Beach (right), near Santa Cruz, California, have been greatly enlarged in recent years; the one at the left is doomed to collapse and leave only an isolated stack. (William Bradley)

sink, but the corals build upward fast enough to maintain their position in shallow water. The fringing reef becomes a barrier reef as the island slowly sinks. When the continuing subsidence finally drowns the summit of the volcanic cone the barrier reef becomes an atoll enclosing a lagoon. A crucial point in Darwin's theory was that the subsidence of the volcanic island was slow enough for the corals to keep pace with it in their upward building of the reef. To support the theory he cited a series of Pacific Islands, many of them in the Marshall and Ellice groups, which suggest a sequence from mountainous islands surrounded by near-shore barrier reefs with narrow lagoons, through less lofty islands surrounded by more distant reefs beyond broad lagoons, to barrier reefs enclosing lagoons containing two or three centrally located, small volcanic islands, and thus at last to true atolls.

The Darwinian theory has been significantly modified in recent years, but subsidence is still an important factor in the more complicated modern reef theory. Detailed knowledge about many atolls in various parts of the Pacific Ocean has been gained by the boring of deep holes and by seismic studies, some of them using atomic bomb tests as the "shot" from which the ground waves are instrumentally recorded. The origin of atolls is far more complex than it first appeared. Their surface similarities conceal striking differences in underground structure. Some are based on truncated volcanic cones. Apparently these were once islands but were eroded completely away, perhaps during times of ice expansion in the Great Ice Age, when sea level was lower and the water so cold that corals could not flourish to protect them from the waves. Where the reef material extends far below the depths in which the reef-building corals can live, subsidence has obviously occurred. If the reef is all shallow, the changes of sea level during Pleistocene time may be sufficient to explain the reef without using local crustal movement as an explanation. In dealing with the rock fragments brought to the surface from boreholes, it is often difficult to differentiate the coral rock, still in place where the corals grew, from the reef material broken loose by waves and re-cemented where it came to rest on lower slopes.

Even so, the subsidence theory, essentially as proposed by Darwin, certainly applies to some of the atolls studied in depth. Bore-holes on Eniwetok Atoll in the Marshall Islands, for example, penetrated downward through coralline limestone to a depth of nearly a mile before reaching the basalt of a volcano whose eroded summit stood two miles above the level of the surrounding ocean floor. Fossils near the base of the limestone cap in one drill hole date back to Late Eocene time and represent shallow-water organisms. This provides a record of over-all subsidence for more than forty million years. It was not continuous subsidence, however; the downward movement was interrupted by several periods of emergence recorded by fossil shells of land animals and the pollen and spores of land plants in zones of recrystallized sediments penetrated by the drill at various depths.

Viewing Eniwetok and the other atolls of the western Pacific in terms of their long geologic history, the picture that emerges is of islands repeatedly rising and sinking. The changes of level were due to both crustal movements and sea level fluctuations during the glacial epoch. Today, some of these coral islands are of great importance in the life of man. The part played by many of them in several crucial battles of World War II will be long remembered. A few now provide essential air bases for trans-Pacific planes.

56. The rugged coast of southernmost Africa has altered little since the sea reached its present level. A bay-head beach can be seen in front of the town of Hout Bay; and famous Table Mountain is visible below the clouds in the distance. (Laurence Lowry: Rapho Guillumette)

9

Glaciers and Ice Sheets

Of all the places that a traveler can easily get to, the Gornergrat in the Swiss Alps affords the most spectacular and informative panorama of glaciers sculpturing the landscape. One takes a cogwheel train that winds up the steep mountainside above Zermatt to the crest of a rugged ridge overlooking the Gorner Valley. There, in good weather, one sees the magnificent snow-blanketed summits of the Pennine Alps stretched in a half-circle from Monte Rosa on the extreme left, through Lyskamm, Castor and Pollux, and other peaks, to the awesome Matterhorn on the far right. Below, in the foreground, the Gorner Glacier moves from left to right, fed by the many tributary glaciers descending from distant snowfields. Here is an ideal observatory for the study of glaciers and their behavior.

Every glacier—whether in the Alps, Alaska, the Canadian Rockies, the Patagonian Andes or the Himalayas—has something to tell the thoughtful observer about the various ways in which slowly moving but immensely powerful streams of ice have shaped and are shaping the landscape (Plates 57 and 60).

Glaciers confined to valleys in mountainous regions are known as *alpine glaciers*. Most of these are *valley glaciers*— long, narrow bodies of ice whose position and movement are largely determined by the local topography. They range in length from less than a mile to scores of miles; the ninety-five-mile-long Hubbard Glacier in the Alaska-Yukon area is the longest valley glacier in temperate regions. Some of the smaller alpine glaciers are wider than they are long. *Piedmont glaciers* are found at the foot of mountain ranges in Alaska and in a few other places where valley glaciers spread out beyond the foothills onto adjacent lowlands; the Malaspina Glacier, for example, covers fourteen hundred square miles between the St. Elias Range, where it is fed by a dozen valley glaciers, and the Alaskan coast.

Icecaps and *continental ice sheets* differ from alpine glaciers in that the ice moves outward in all directions from centers of accumulation; shape and movement are influenced but not determined by the configuration of the land beneath them. Such icecaps as the Jostedalsbree on a high plateau in the interior of Norway and the Vatnajokull on an upland in Iceland are good examples. At places around their margins, tongues of ice, like the Svarti-sen Glacier, may extend into a valley leading downward from the edge of the plateau. Continental ice sheets are similar but are much more extensive, covering a large part of a continent. The only remaining ones are in Greenland and Antarctica, but there have been many others in the past.

Barnard Glacier, Alaska, displays an amazing pattern of moraines. In the distance the lateral moraines on the farther sides of two tributary glaciers unite in a medial moraine that runs down the center of the photograph. Three smaller glaciers flow in from the right, each adding another medial moraine.
(Bradford Washburn)

154

Regardless of shape or movement, all bodies of glacial ice have much in common. Most significantly, glacial ice originates from snow; it is not frozen water, although a small fraction of the ice in the lower part of most glaciers is frozen meltwater. The transformation of snow into glacial ice begins when the snow is compacted. This may result merely from the settling of the lower part of a snowbank under the weight of accumulating snow or from the packing of snow by wind. In either case the fluffy snowflakes recrystallize into denser granules that, when partially consolidated, are known as *firn* or *névé* (the terms are practically synonymous). This re-organization of molecules is as much the result of time as of pressure. The snow that lingers into April or May in a shaded nook is heavier and more granular than it was when it fell in January or February. Indeed, the word *firn* is derived from a German adjective that means "of last year." With further compacting and recrystallization as new snow is added, the granules of firn are welded together into ice. Air pockets and water bubbles are responsible for the blue color of the clean ice in every glacier (Plate 58).

At low temperatures and under slight pressure, glacial ice is brittle, but near its melting temperature and under greater pressure it deforms plastically without cracking or breaking. In most glaciers outside the Arctic and Antarctic regions, the ice is near the pressure melting point most of the time, except for the shallow surface layer in which there is recurring winter chill. Such ice becomes plastic under pressures equivalent to the weight of a column of ice only a couple of hundred feet high.

The shape of every alpine glacier and its relation to the surrounding topography suggest immediately that glaciers flow down valleys. That is true only if by flow we mean the "solid flow" of a plastic substance, not the "liquid flow" of a fluid. Solid flow takes place in the deeper interior of the glacier where pressure makes the ice plastic. The ice near the surface of the glacier remains brittle; carried along on the slowly flowing ice, it is likely to be broken by any change in slope of the glacier's bed. Hence deep crevasses appear on the surface and near the edges of the moving ice. In temperate latitudes these cracks are never more than a hundred feet deep, whereas on the Antarctic ice sheet, where the ice remains colder throughout the year, they are often twice that depth. It requires greater pressure, and hence a thicker overburden, to transform brittle ice into plastic ice at the lower temperature.

Blocks of rock are held firmly in the brittle ice at the thin edges and front of a glacier, and there they serve as graving tools to scratch and gouge its bed. Even when such tools are embedded in potentially plastic ice near the bottom of the thicker part of a glacier, they are often held rigidly enough to erode the bed as they are pushed along. This use of such tools to gouge out hollows in solid rock helps to explain the differences between erosion by glaciers and by rivers. The movement of ice as a result of plastic flow is generally greater in the higher parts of most alpine glaciers than farther down their courses. Where glaciers are well nourished by abundant snowfall in their areas of accumulation, much of the movement below snow line results from sliding, not plastic flow, over the rock bed. Such glaciers are very effective agents of erosion.

The rate and nature of movement of glaciers vary greatly and the conditions determining them are complex. As a rule it is helpful to distinguish between those parts of a system of glaciers in which the volume of ice is increasing as a result of snow accumulation and other parts where the ice is wasting away as a result of melting and evaporation. Loss of ice is known as *ablation*. The transition from areas of accumulation in the higher parts of a glacial system to areas of ablation in its lower parts is marked by significant changes, both in the appearance of the glacier and in the stresses in its interior. The relationship between accumulation and ablation, the slope and shape of the glacier's bed, and the temperature of the ice are the prime factors in determining its movement.

Measurements on many glaciers indicate that the rate of ice movement is generally between one hundred and three hundred feet per year, although in a few glaciers it is much slower and in others more rapid. Ordinarily, the thicker ice in the middle of a valley glacier moves more rapidly than that at its sides, the latter being retarded by friction against its bed. On some glaciers there are curving bands of dirty ice which alternate with bands of cleaner

The extraordinarily large Malaspina Glacier at the foot of Mt. St. Elias, Alaska. Where it impinges against Marvine Glacier (at far right), its bands of debris-laden ice form a zigzag pattern. (Bradford Washburn)

ice. The dirt in the ice may have been wind-blown dust that fell on the snow during the summer when adjacent slopes and peaks were bare; the clean ice was formed in winter when the mountain range was blanketed in snow. When this is the origin of the bands, it is possible to determine the number of years required for the accumulation of the ice in that part of a glacier.

The pattern of movement in a glacier is usually revealed by lines or ridges of debris on its surface. These are known as *moraines—lateral* if they accumulate along the sides of the glacier, and *medial* if formed by the merging of lateral moraines where tributary glaciers join to make a trunk glacier, like the one in the Gorner Valley below the Gornergrat. The many medial moraines on such glaciers as Barnard Glacier in Alaska's St. Elias Mountains are especially spectacular and give a strong impression of the mobility of glacial ice—mobility with little or no mixing of ice from different sources. An equally impressive but quite different pattern is displayed on certain piedmont glaciers, such as the Malaspina, where the varying rate of movement of the ice from many valley glaciers, coalesced beyond the mountain front, produces an amazing set of moraines that curve and re-curve in intricate fashion.

GLACIAL EROSION

Although glaciers differ profoundly from rivers in their movement and therefore in their work as agents of geologic change, they resemble rivers in having regions of erosion, transportation and deposition. Erosion is especially active in the high basins at the heads of alpine glaciers. Where these basins are rimmed by a semicircular wall of steep cliffs, they are known as *cirques*. Commonly a deep, wide crevasse with an overhanging upper lip marks the beginning of the movement of accumulating firn-ice. In this so-called *bergschrund* (the German word meaning "mountain fissure"), frost action is especially strong. The fragments of rock loosened by frost from the foot of the cirque's headwall are embedded in the ice for transport. Meanwhile, wedges of ice penetrate into the cracks opened in the bedrock, and large blocks of rock are plucked away by the moving glacier. Since glacial erosion is thus especially effective in the basins of accumulation, it is not surprising that cirques, with their steep headwalls and semicircular ground plan, are the hallmarks of alpine glaciation.

Among the other tools of erosion are the rock fragments loosened by weathering from the cliffs above the glacier. These fall, slide, or are swept by avalanches of snow into the bergschrund or onto the glacier just below it. The larger angular blocks gouge and scratch the bedrock, forming grooves or scratches called *striae;* the finer fragments smooth and polish the rock floor beneath the moving ice. Thus the autograph of a glacier may remain on its former bed long after it has dwindled or disappeared altogether. Moreover, many rock fragments in transit within the ice are similarly marked. Stones held in the base of the ice are ground and scratched as they erode the glacier's bed. Those in the midst of the ice are occasionally worn by contact with each other as parts of the ice body shear past or over other parts. Since the stones sometimes change position when the ice temporarily relaxes its grip, projecting angles are rounded off and two or more sides of a

stone may be flattened, smoothed and scratched. Striations on such stones run in many directions. Their final shape is notably different from that of pebbles, cobbles, and boulders shaped by running water, waves, or wind. The composition of the stones carried and deposited by most glaciers is extremely variable—granite, limestone, quartzite and gneiss all mixed together, for example; they are appropriately called *erratics*.

Erosion is also vigorous in the outlet valleys leading downward from the névé fields in basins of accumulation; this is especially true in the beds of trunk glaciers, such as the Gorner Glacier, that are nourished by ice from many tributaries. The rock floor across which the ice moves is deepened and scoured, and spurs between tributaries are snubbed away. The resulting deep U-shaped valleys (as seen in a lengthwise view) are appropriately known as *glacial troughs;* the valley below Garmisch in the Bavarian Alps is a good example (Plate 64). In general, the amount of valley deepening is definitely related to the size of the glacier; tributary valleys, with their smaller glaciers, are not eroded as deeply as trunk valleys with their larger glaciers.

Three coalescing glaciers cause contortions in lateral and medial moraines in the Sisitna Glacier, Hayes Range, Alaska. (Bradford Washburn)

159

Consequently, when the ice disappears, the floors of tributary valleys are left high above the floor of the main valley; they are *hanging valleys,* quite different from the normal tributary valleys of a landscape sculptured only by running water.

These effects of glacial erosion quickly catch the eye in many mountain regions that are now or were once occupied by vigorously eroding alpine glaciers. High on the mountainsides, but not actually at their summits, are the basins in which snow accumulates to become ice. The steep headwalls of cirques are easily recognized, whether or not the basins are filled with ice. Between the small glacial valleys leading downward from adjacent cirques are sharp, jagged ridgecrests known as *arêtes,* often lined with small pinnacles of rock called *gendarmes* by the French-speaking alpinists. Deep U-shaped troughs mark the courses of the main glaciers; the spurs of rock that ordinarily project into stream-carved valleys between tributaries have been snubbed away, leaving triangular facets along their sides. From hanging valleys, streams cascade down steep slopes to join the main rivers in deep-scoured glacial troughs. Bridal Veil Falls (Plate 65) in Yosemite National Park, California, for example, is the result of differential erosion by a trunk glacier in the Merced Valley and a tributary glacier which joined it at that place. And many of the beautiful falls in Norway plunge into fjords from hanging valleys high on the sides of those glacial troughs; among them are the famous Seven Sisters Falls in Geirangerfjord and the spectacular "foss" in Sognefjord near Balheim.

For glaciers, there is no profile of equilibrium as there is for rivers. Moving ice can gouge out great hollows where rocks are weak or where the working tools are unusually abundant or are applied with special vigor. Such rock-rimmed basins on valley floors, when filled with water after the ice has disappeared, are called *tarns*. They are commonly found on cirque floors or along the bottom of the smaller glacial troughs.

The concept of an erosion cycle, with youth, maturity, and old age, applies to glaciers as to rivers. In mountains where glaciers were never large or were present for only a short time, the flanks of rounded, stream-worn summits may be marked by small cirques, each with its characteristic headwall. Where extensive unglaciated summits remain, it is evident that glacial erosion stopped in the youthful stage of the cycle. One can easily imagine what would have happened if glacial erosion had been more vigorous or had continued longer. The headwalls of the several cirques would have been eroded back until only a jagged peak was left where the rounded summit had been. This is the typical sequence in the sculpturing of such a maturely glaciated peak as the monumental Matterhorn (Plate 62), the beloved Mont Cervin of French-speaking Swiss. Any peak thus carved by the headward erosion of alpine glaciers may be described as a *matterhorn peak,* but few indeed are as magnificently sculptured as the one that gives its name to the class.

GLACIERS AS CARGO CARRIERS

Wherever the ice of a glacier is moving, whether by solid flow or by sliding, it is a transporting agent. The cargo of rock fragments may be carried in the

bottom ice, in the midst of the ice, or on the surface of the glacier. Much of it is concealed from view but enough is visible, especially in summer, to indicate the great power of glacial ice as an agent of both erosion and transportation. On most alpine glaciers, the debris in transport is largely concentrated in the medial and lateral moraines. Some of these are actually ridges of ice thickly veneered with rock fragments which protect the ice from melting as rapidly as elsewhere on the glacier's surface. Some lateral moraines rest on the bedrock just beyond the edge of the glacier, indicating a diminution in the volume of ice during recent centuries.

Melting and evaporation of glacial ice below the snow line depend primarily on air temperature and sunshine. The rate of ablation therefore fluctuates from season to season and hour to hour. Rock fragments on the ice also affect the rate of melting and are responsible for many of the surface features peculiar to glaciers. Dark rock fragments absorb the heat of solar radiation but clean ice reflects it. Dust particles, sand grains and small pebbles warmed by the sun melt the adjacent ice; the surface of dirty ice is lowered much more rapidly than that of clean ice. Small chunks of rock, three or four inches across and an inch or so thick, transmit the sun's heat from their upper to their lower surfaces and melt their way down in the ice to form pits or "wells" sometimes several inches deep. Larger rocks, too thick to conduct heat effectively from upper to lower surfaces, shade the ice on which they rest and may eventually stand on pedestals when the surround-

Partly undercut by the ice of the Mer de Glace, the spectacular cliffs of the Aiguille du Plan and Blaitière in the French Alps have weathered rapidly, leaving jagged needles and deep clefts. (Bradford Washburn)

161

ing ice has melted away. Huge slabs of rock, perched on ice pedestals a few feet high, may be seen on some alpine glaciers. They are called *glacier tables* and serve as vivid indicators of the amount of wasting away in recent years.

GLACIAL DEPOSITS

The lower end or snout of some valley glaciers is so thickly and continuously veneered with rock fragments that it is impossible to locate the actual terminus of the ice. More commonly, however, a well-defined *frontal moraine* or *end moraine* loops around the glacier's snout from the lateral moraine on one side to that on the other. Beyond the present terminus of many glaciers, ridges or irregular mounds of boulder-studded debris form loops across the valley floor. They were frontal moraines when the glacier was longer than it is today. Similarly, on valley walls, well above the sides of many glaciers, are more or less continuous ridges of debris that obviously are lateral moraines deposited when the glacier was wider.

To produce conspicuous frontal and lateral moraines, there must be a delicate balance between the nourishment and wastage of a glacier for a considerable number of years. Whatever the rate of forward movement of the ice, it must be approximately equalled by the rate of melting at the end and sides of the glacier. Thus each year the load of rock debris carried in or on the advancing ice is dropped in approximately the same position; the glacier as a whole neither advances nor retreats. Then, if the glacier should retreat, the frontal moraine that marks its maximum extent is called a *terminal moraine*.

MELTWATER

Countless rills of meltwater trickle over the surface of the lower part of many valley glaciers during warm days in summer when the ice melts most rapidly. The rills unite in streams that cut channels in the ice and carry off debris in accordance with the same factors that regulate the work of running water. Many of them descend into crevasses and flow onward within the glacier or beneath it. Walking across a glacier, one may sometimes hear the gurgle of an invisible stream in the ice. Occasionally a surface stream plunges with a mighty roar into a more or less circular hole in the glacier. Swiss and French alpinists call this a *moulin* (mill) because of the noise made by the rushing water; and the word has been adopted for such holes in the ice whether or not the water makes enough noise to resemble a mill. If the ice remains nearly stationary for a sufficient time, the plunging water, well supplied with pebbles and cobblestones, may deepen the moulin clear through the ice into the rock floor on which it rests. There a pothole may be worn deep into the bedrock by the stones swirled around in the turbulent water. In this way the remarkable pits in the Gletschergarten at Lucerne, Switzerland, were formed when ice occupied the valley.

The meltwater in and beneath a glacier finally emerges at its snout, there to mingle with the water melting from the ice front. Sizable tunnels are

formed in some glaciers by the subglacial streams; to explore the interior of such a tunnel is a thrilling experience.

Meltwater from a glacier is almost always loaded to capacity with detritus from the start. Glacier-fed streams are usually gray with "rock-flour," pulverized by the grinding action of sliding and shearing ice, and held in suspension in the turbulent water. Particles of sand-grain size may ride in suspension for long distances if the streams are swift, whereas pebbles and cobblestones are rolled and bounced along the stream beds and dropped at various distances from the ice front. Thus *glaciofluvial deposits,* largely pebbles and cobblestones, begin to form immediately beyond the frontal moraines. Farther down the valley they generally include sand mixed with gravel. If a stream in a youthful valley derives much of its water from melting glaciers it will build up a valley flat resembling the floodplain of a postmature river. Such glaciofluvial deposits, known as *valley trains,* may extend for many miles down valleys.

The Great Aletsch Glacier moves down a gigantic glacial trough from the flanks of Jungfrau and Monch in the Swiss Alps. Its ice has dammed the tributary valley in the lower right to create the lake known as Marjelensee. (Photo Klopfenstein)

GLACIERS AND CLIMATE

The study of successive frontal moraines downvalley from an alpine glacier provides trustworthy clues to the nature and timing of past climatic changes. The age of moraines is indicated by the character of the vegetation that has gained a foothold on them. The moraines now building up at the edge of the ice are generally without any plant life except moss and lichens. The next older ones may have only grass and shrubs, and the next may support small saplings, whereas still older moraines may be occupied by trees of varying size whose age may be determined by counting their growth rings.

It now appears that few if any of the glaciers in the Alps, the Rockies and

163

other mountain ranges in temperate and tropical latitudes are, as was formerly assumed, the dwindling remnants of the much larger glaciers of the Great Ice Age. Ten thousand years ago the glaciers were retreating and by 3000 B. C. most if not all of them had completely disappeared from those mountains. That was the time of a world-wide "Climatic Optimum" or "Thermal Maximum." The last millennium B. C. was marked by a return toward harsher climatic conditions; there was a widespread recrudescence of glaciers and the "Little Ice Age" began. With some slight warming of the climate between A. D. 500 and 1000, this has continued almost to the present. Most existing glaciers, born early in the Little Ice Age, reached their maximum size in the latter half of the eighteenth or early half of the nineteenth century and have been retreating ever since. Some of the largest, high altitude, trunk glaciers continued to advance even in the first decade of the twentieth century but most of them have now joined the retreat. Intermittent recession of mountain glaciers during the last hundred years seems to be the general rule. The acute sensitivity of glaciers to climatic change is demonstrated by as many as thirty recessional moraines within a few miles of the present terminus of certain glaciers in Alaska, Scandinavia, and Patagonia, all left behind within the last two hundred years. We are still in the shadow of the Little Ice Age, but the current trend is toward a return to the conditions of the Climatic Optimum of five thousand years ago.

Studying the glaciers of the Little Ice Age is good preparation for understanding the features resulting from the action of the larger alpine glaciers during the Great Ice Age. Some effects of the larger glaciers have already been noted. The Matterhorn, for example, could hardly have been sculptured to its present shape by the relatively small and short-lived glaciers of the Little Ice Age. The ice responsible for its towering lineaments was that of earlier stages; the existing glaciers, effective though they are, merely added the finishing touches to a job largely completed before they came into existence. The same is true of the magnificent cirques in all vigorously glaciated mountains; they were carved by earlier ice and then re-occupied by the modern glaciers. In much the same way, the long glacial troughs, such as the one extending many miles downstream from Chamonix, were sculptured long before such glaciers as the Mer de Glace came into existence.

ICECAPS AND CONTINENTAL ICE SHEETS

Less than 4 per cent of the earth's glacial ice is in alpine glaciers today; more than 96 per cent is in the vast continental ice sheets of Antarctica and Greenland and the scores of icecaps on plateaus or highlands, many of which are in far northern latitudes. Although the movement of this ice is strongly influenced in many places by the topography under the glacier, this is not the controlling factor that it is for alpine glaciers. Instead, the ice spreads outward under its own weight like a blob of tar on a flat surface. In the mass movement of icecaps and ice sheets, solid flow predominates over sliding, just the reverse of their importance in alpine glaciers. Movement by sliding, however, may be dominant for the thinner and more rigid ice near the edges of the cap or sheet as the central plastic mass shoves it outward.

THE GREENLAND ICE SHEET

Ice covers all of the interior of Greenland, an area of more than seven hundred thousand square miles. The surface of this ice sheet has the profile of a shield rising gently from its margin to form three broad domes, the highest more than ten thousand feet above sea level. Geophysical measurements indicate that at some places the ice is more than seven thousand feet thick. It pushes outward against the mountains that rise along the east and west coasts and sends tongues of ice through valleys to the sea. In certain of these outlet glaciers the velocity of ice movement exceeds a hundred feet per day.

The rock floor beneath the Greenland ice is an elongate oval saucer, its central portion approximately at sea level and its rim rising to altitudes of five thousand feet or more in the peaks and ridges of the coastal mountains. Some peaks as far as forty miles inside the ice margin are completely surrounded by ice not quite thick enough to top them. Such hills of rock in the midst of the ice are known as *nunataks*. Glacial striae and ice-sculptured surfaces on the flanks and summits of these nunataks indicate that the ice sheet was formerly much thicker than it is today. It was also more extensive, as indicated by glaciated rock surfaces and morainal debris far beyond the present ice margin. Many of the outlet glaciers terminate in the middle of magnificently sculptured glacial troughs; they must formerly have extended the full length of those troughs. Many troughs have been partially drowned by the post-glacial rise of sea level and constitute the spectacular fjords of the Greenland coasts.

All the known facts indicate that the Greenland ice sheet is a holdover from the Great Ice Age. Expansion in the Little Ice Age, beginning only a few thousand years ago, has been followed by appreciable diminution. During the present century the ice front has been retreating, or at some places has remained stationary.

Meltwater from a receding glacier in McKinley National Park, Alaska, continues to deposit a valley train, the sand and gravel of which have here been interrupted by remnants of a moraine marking the former ice front.
(Bradford Washburn)

Surprisingly enough, the only extensive flat land of Greenland not now covered by ice is in the far north. Much of this region, known as Peary Land, is an undulating plain or low-lying plateau. Its northern shore is the land nearest the North Pole. Beneath the tundra the ground is frozen the year around—a state called *permafrost*. But it is beyond the present limits of the ice sheet and much of it almost certainly was never covered by glacial ice even during maximum refrigeration in the Great Ice Age. The reason for this anomaly is not hard to find. To produce a great ice sheet, the climate must be both cold and wet; the climate of northern Greenland is cold enough but not wet enough. Snowfall at Thule is sufficient to meet the needs of Air Force personnel for winter sports but it is only a tenth as much as on the southeast and southwest coasts. Without abundant snowfall, this northern sector of the ice sheet is so poorly nourished that its margin has been practically stationary for centuries. This is a fortunate circumstance. Difficult though it is to maintain an airport for jet-propelled planes in a region of permafrost, it would be almost impossible to do so on the surface of an ice sheet. And Thule is as essential for the "polar flights" of commercial transport lines as it is for any military purpose.

THE ANTARCTIC ICE SHEET

The ice sheet on Antarctica covers more than five million square miles and contains approximately nine-tenths of all the glacial ice on our planet. Most of it is more than a mile thick and at several places the thickness exceeds ten thousand feet. It completely buries mountain ranges that stand thousands of feet above the surrounding land. The peaks and ridges of still loftier mountains rise above its surface as nunataks similar to those on Greenland. Some, however, are so large that they have catchment basins on their flanks from which glaciers move outward to merge with the surrounding ice.

At several places around the edge of the Antarctic Ice Sheet, the ice has pushed outward into water deep enough to float it, producing an ice shelf. The largest ice shelf covers half of the Ross Sea, a great bay opening outward toward New Zealand and the western Pacific Ocean. The area of the Ross Ice Shelf almost equals that of Texas, but only its seaward part is afloat. Its ice is a thousand or more feet in thickness and moves outward as a unit to terminate in a glistening cliff from which huge icebergs break away and drift far to the north (Plate 61).

Elsewhere there are ice-free areas of land beyond the front of the ice sheet. One of the best-known of these is at the head of McMurdo Sound, just to the west of the Ross Ice Shelf. Here, lobes and tongues of ice extend into valleys where moraines mark the former positions of the glacial snouts. Moraines nearest them are actually ridges of stagnant ice thickly veneered with glacial debris. Radio-carbon dating of algae from extinct ponds in these ice-cored moraines indicate that they are at least six thousand years old.

Such data bear upon one of the unsolved problems of the Antarctic Ice Sheet. Is it now so well nourished by the annual accumulation of snow that it is growing larger year by year? Or are the present climatic conditions such that it is slowly dwindling in its over-all dimensions? Estimates of recent annual increments of snow, compared with estimates of annual loss by

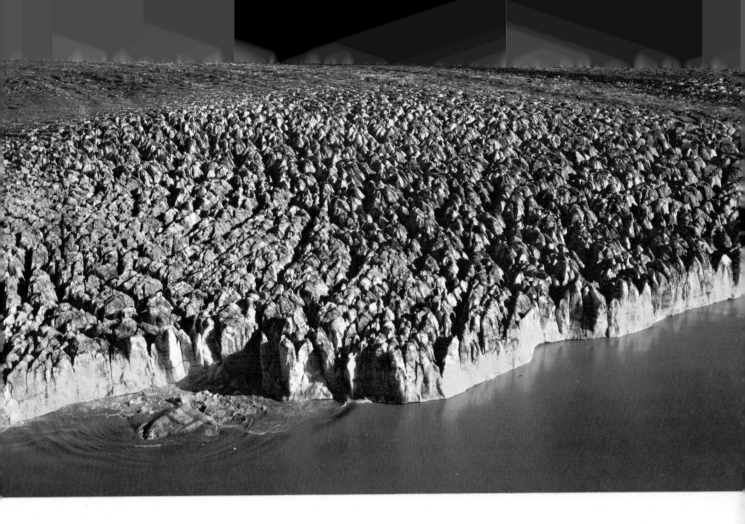

ablation and calving (breaking off) of icebergs seem to indicate that the ice sheet is growing larger, but the data are not yet conclusive. Observations in the McMurdo Sound region suggest the opposite. It is reasonable to think that the Antarctic Ice Sheet has responded to worldwide climatic change in much the same way as have the glaciers of the Alps, the Rockies, the New Zealand Alps and other mountain ranges. It attained maximum dimensions some thirty thousand years ago when ice was piled against certain nunataks to a height of eight hundred feet above its present surface. During the Climatic Optimum of five thousand years ago, it dwindled to dimensions only a little less than those of today, then expanded slightly during the Little Ice Age, and then decreased somewhat in its dimensions.

The front of Miles Glacier. Alaska, while "calving" several icebergs. (Bradford Washburn)

GLACIATION IN THE GREAT ICE AGE

Almost every mountain throughout the world that rises a few hundred feet or more above the local timberline was sculptured by alpine glaciers during the Great Ice Age. Glacial cirques, matterhorn peaks, jagged arêtes, tarns, ice-scoured troughs, and hanging valleys reveal the former presence of those glaciers. These are the features that give many mountains their grandeur and spectacular beauty.

The altitude of timberline and of the usually somewhat higher snow line depends upon many factors. Most important is the latitude. In general, timberline is highest in equatorial latitudes and is at progressively lower elevations toward either pole. Its position, however, is strongly influenced by the circulation of the atmosphere. The movement of air masses may bring abundant moisture to one side of a mountain range and leave the other side comparatively dry. Some mountains are in arid regions, others in humid regions; some are buffeted by strong winds, others are rarely so affected. Moreover, there is exposure to sunlight: north-facing slopes in northern latitudes are shaded from the sun whereas the south-facing slopes are not. All of these factors influence the accumulation of snow and its transformation into glacial ice. For alpine glaciers to form, much more snow must fall in winter than melts away in summer; almost certainly, there will be winds to sweep the snow from exposed slopes into sheltered hollows or basins. With notable differences between mountain ranges in the same latitude, it follows that one must climb to fifteen thousand feet in the Ecuadorean Andes to find results of glacial erosion and to eight thousand feet in the Colorado Rockies, whereas in Alaska and Scandinavia they appear at sea level.

Extensive areas of the earth's surface—in the northern United States and in Canada, in northwestern Europe and in Patagonia, for instance—are liberally strewn with great boulders and irregular piles of gravel, sand and clay quite unlike anything observable elsewhere. The boulders may consist of rocks wholly different from the underlying bedrock. Thus, loose blocks of granite weighing many tons and apparently identical with the granite of central Ontario rest on the sandstone and limestone strata of southern New York, central Ohio, and central Illinois, hundreds of miles from the ledges of granite whence they must have come. Similarly, other boulders, composed of one or more of the igneous or metamorphic rock types of the Scandinavian highlands are found on the sedimentary rocks of the Netherlands and the North German plain, hundreds of miles from their source.

In these areas, mounds or hillocks of sandy clay containing cobblestones, blocks and boulders of a great variety of rocks, are scattered over hills and valleys. The landscape is dotted by scores of lakes and ponds haphazardly distributed. Occasional ledges of outcropping bedrock are smoothed and polished, or grooved and striated, like the bedrock around the edges of dwindling valley glaciers in the Alps. It is not surprising that in 1840 Louis Agassiz, then Professor of Natural History at the University of Neuchâtel in Switzerland and fresh from studies of nearby glaciers, espoused the audacious theory that "at a period geologically very recent, the entire hemisphere north of the thirty-fifth parallel had been covered by a sheet of ice possessing all the characteristics of the existing glaciers in the Swiss Alps." The glacial theory met strong opposition from many scientists at the time but it gradually gained adherents and by 1880 it had become one of the accepted principles of geology.

Ideas concerning the ice sheets of the Great Ice Age have been radically altered, however, since the time of Agassiz. We now know that, instead of one great polar icecap, there were several separate ice sheets. In North America the two largest had their centers in Canada, one to the east and the other to the west of Hudson Bay. In each of those areas the snow accumulated, winter after winter, century after century, until the weight of the pile caused the lower portion to recrystallize into ice. The ice increased in amount until it expanded over the land, moving outward in all directions.

In Europe, the largest ice sheet originated on the highlands of Scandinavia.

57. The Morteratsch Glacier near the Swiss-Austrian border is fed by snow from basins in distant mountains. At lower altitudes the melting ice has deposited moraines around the glacier, while a medial moraine covers part of the crevassed ice. (Beringer and Pampaluchi)

58. An ice cliff on the Knik Glacier (above, left), Alaska, shows horizontal bands from dust gathered on the snowfields at the glacier's source. (Steve McCutcheon)

59. Meltwater on the Ward Hunt Ice Shelf (left), Ellesmere Land, has cut a channel in the ice. A large part of the ice shelf has since broken loose and become a "floating island." (John R. T. Molholm)

60. The Taschorn and Dom (above) stand high above the seracs and crevasses of a glacier at Valais, Switzerland. (L. Gensetter)

61. The towering wall of ice edging the Ross Ice Shelf (right), Antarctica, shows the bands resulting from seasonal climatic change. (John R. T. Molholm)

62. The Matterhorn in the Swiss Alps is the archetype of an excessively glaciated mountain peak. (Bradford Washburn)

63. Only a few small glaciers remain in the Pyrenees today, but this U-shaped valley (above, right) leading down from the snowfields was deeply eroded by a large glacier during the Great Ice Age. (Michel Terrasse)

64. Valleys such as this in the Tyrolean Alps (below, right) are "glacial troughs," formerly occupied by vigorously-eroding trunk glaciers. (Gerhard Klammet)

65. During the Great Ice Age the Yosemite Valley, California, was eroded so deeply by the vast Merced Glacier that a tributary valley, together with magnificent Bridal Veil Falls (right), was left "hanging" high above its floor when the glacier disappeared. (Robert Clemenz)

66. For scores of miles upstream from its junction with the St. Lawrence (in the distance), the Saguenay River (below) is in a picturesque, steep-walled valley—the Canadian counterpart of a Norwegian fjord.
(Laurence Lowry: Rapho Guillumette)

67. In the foreground of this view of the Athabaska River and Mt. Athabaska (facing page) in the Canadian Rockies is a huge glacial moraine forming a dam through which the river has recently cut its way.
(Andreas Feininger)

68. Along the edge of the Antarctic Ice Sheet
an ice cavern with icy stalactites and
stalagmites duplicates the features in limestone
caves. (Robert Nichols)

Expanding southward and eastward, it reached far into Germany, Poland and the Soviet Union. Toward the southwest and west it crossed the floor of the North Sea and for a time coalesced with local icecaps on the highlands of Scotland, Wales and England. Thus glacial debris north of London includes boulders transported by ice from Norway.

Furthermore, we now know that the Great Ice Age had a very complicated history, involving the formation, advance and retreat of successive ice sheets, each of which occupied much the same territory. At four different times the ice moved outward from the major centers of accumulation in North America and Europe. Four times the climatic pendulum reversed its swing; the temperature of the ice-covered lands gradually rose until the ice melted away, leaving its load of transported boulders and pulverized rock strewn over the land. The four glacial stages and the three interglacial intervals constitute the Pleistocene Epoch of the Quaternary Period (see the Geological Timetable, page 312).

During at least one of those intervals, the climate of North America was warmer than it is now. Osage oranges, pawpaws and locust trees grew near Toronto, well beyond their present northern limit. Fossil leaves of these and other warm-climate plants are preserved in beds of clay and sand between bouldery clay of an earlier glacial episode and the deposits left by the latest ice sheet. Although there is no way to measure precisely the length of these interglacial intervals, it is quite certain that even the briefest of them lasted several times as long as the time that has elapsed since the latest ice began to retreat from its maximum spread.

This history of the Great Ice Age is deduced from detailed studies of the erosion and the deposits of successive ice sheets in many localities. All the materials transported by glacial ice and deposited by it or its meltwater are known collectively as *glacial drift*. Unsorted drift is *till*; in it, boulders and pebbles, large and small, are embedded at random in a matrix of clay or a mixture of sand and clay. The microscope reveals that this matrix is pulverized rock rather than the product of rock weathering. Had it been transported by running water, it would have been sorted by size and weight. Many of the stones in till have irregularly rounded surfaces or smoothly planed faces marked by parallel scratches and grooves, the result of the grinding action of stone against stone and stone against rock floor as they moved along in the grip of the ice.

Till commonly occurs in moraines. The frontal moraines left by a continental ice sheet are usually well-defined ridges or zones of rounded hillocks with intervening hollows—the "knob and kettle" topography of certain regions. The moraine that marks the farthest reach of an ice sheet is a *terminal moraine;* those that indicate temporary halts in the retreat of the ice front are *recessional moraines*. The recession of the last ice sheets involved several re-advances subsequent to major retreats. Thus it is possible to recognize the substages set forth in the table on page 312. Patches of till are likely to be found anywhere within an area formerly covered by ice. Where fairly thick and extensive, and especially where they give the landscape a hummocky appearance, they form *ground moraines*. These are irregularly distributed and may have been deposited under "live" ice far within its margin, or may have been left by stagnant ice that melted downward from its surface rather than backward from its edges.

The glacial till of a moraine in the Val d'Herens, Valais, Switzerland, has been carved by rain into pyramids and pinnacles, some of them capped by large erratics. (Photo Klopfenstein)

Till is also the major component of *drumlins*. These are smoothly rounded, elongate oval hills that generally occur in groups, with the longer axis of each hill parallel to that of its neighbors. They are conspicuous features of the landscape in eastern Massachusetts, including Bunker Hill and Breed's Hill of Revolutionary War fame, as well as in New York State south of Lake Ontario, in Michigan south of Charlevoix, and in southern Wisconsin. Drumlins are especially numerous in Estonia (now a part of the Soviet Union) between the Gulf of Riga and Lake Peipus, where they occur in groups or clusters whose patterns are similar to those in Massachusetts or Wisconsin. In all places the alignment of these oval hills is in the direction of the ice movement.

Glacial debris, transported and deposited by water after its liberation from the ice, comprises *glaciofluvial, glaciolacustrine,* or *glaciomarine* deposits, depending upon the nature of the water responsible for its deposition—streams, lakes or seas.

In any moraine, whether left behind by a continental ice sheet or an alpine glacier, there may be lenses or wedges of stratified sand and gravel in the midst of the till. In some localities these are responsible for shallow artesian wells so beneficial to the rural inhabitants of certain glaciated regions. Fringing the outer margin of many terminal and recessional moraines in the North German Plain, the upper Mississippi Valley and elsewhere are *outwash plains,* built up by countless streams of meltwater that spread deposits of silt, sand and gravel beyond the ice front in a fashion quite similar to that of a piedmont alluvial plain. Likewise, *valley trains* may extend far down the courses of major rivers whose headwaters are fed by the melting ice of a continental ice sheet, even as in an alpine valley.

When certain ice sheets, such as those that last covered Finland and New England, were in retreat, the ice remained stagnant for many hundred years. The waning of those great glaciers resulted as much from decreasing nourishment in the regions of accumulating snow as from rising temperature near their margins. Thousands of square miles of stagnant ice melted downward from the surface; lobes of ice remained in valleys long after adjacent hills were bared. Streams of meltwater deposited torrential cones and alluvial fans of glaciofluvial debris at the edges of great slabs and blocks of the stagnant ice. When it finally melted away, removing its support of the apex of cone or fan, the sand and gravel slumped downward. Thus were formed the rounded hillocks and mounds of irregularly stratified sand and gravel known as *kames*.

At the same time, other streams of meltwater were flowing in ice-walled clefts on the surface of the "dead" ice or in tunnels beneath it. These deposited part of their load along their channels, piling it high between the supporting ice walls on either side. When the ice finally disappeared, these deposits were left as long, smooth-topped, rounded ridges of irregularly bedded sand and gravel, a few feet to one or two hundred feet high and from a few hundred yards to scores of miles in length; these are known as *eskers*. Thousands of eskers in Scandinavia, such as the amazing Punkaharju esker in Finland, in Maine, Wisconsin and elsewhere, wind across the landscape like abandoned railroad embankments.

Kames and eskers are excellent sources of sand and gravel for construction of buildings and highways. In many urban and suburban areas, some of them have already been completely removed. Elsewhere, within the glaciated regions of Europe and North America, numerous gravel pits are being extended in these glaciofluvial deposits.

Many ponds and small lakes, beyond the edge of the retreating ice or among the blocks and slabs of stagnant ice, were occupied by glaciolacustrine deposits carried into the standing water by meltwater streams. These deposits are generally irregularly bedded sand and fine gravel, sometimes resembling deltas. Where ponds were completely filled, sand plains were produced. These, too, are excellent sources of materials for construction. One of them was responsible for the "Desert of Maine" described on page 84.

Where the meltwater trickled slowly into ice-front lakes or where more distant lakes were entered only by streams with gentle gradients, the glaciolacustrine deposits are clay and fine silt rather than sand and gravel. At many places in Scandinavia and New England, conspicuously laminated clays were deposited on lake bottoms. The coarser layers were deposited in the warm season each year, the finer layers in the winter when the lake was frozen over and the currents that tend to hold fine clay particles in suspension were reduced. Each pair of layers, called a *varve* (from the Swedish *varv*, "a periodic repetition"), represents a year. Many varved clays have been studied in minute detail in Sweden where the chronology extends back, with one gap, from A. D. 1900 to about 15,000 B. C. Similar studies in the Connecticut Valley indicate that approximately four thousand years were involved in the retreat of the ice front from Hartford, Connecticut, to St. Johnsbury, Vermont, a distance of about two hundred miles.

Glaciomarine clays are not varved; in salt water the flocculation of ultra-fine particles of clay causes them to settle as rapidly as the coarser particles and the deposit is massive rather than laminated. Even so, it is just as suitable for the manufacture of brick and tile as is the varved clay of an ice-front lake.

GREAT LAKES AND INLAND SEAS

The history of the five North American Great Lakes is intimately associated with the recession of the last ice sheet. At that time, northeastern North America stood lower with respect to sea level than it does today and a large bay extended from the Atlantic Ocean into the Lake Ontario basin. As the continental glacier shrank and eventually disappeared, the land tilted upward toward its present altitude, presumably as a result of isostatic adjustment to the removal of the weight of ice. Thus the lakes were finally brought to their present geographic relations.

Similarly, the changing geography of the Baltic region in Europe during Late Glacial and Postglacial epochs is associated with the recession of the Scandinavian Ice Sheet. During the withdrawal of that ice from its farthest reaches in Denmark and Germany, the fresh water of the "Baltic Ice Lake" occupied the southeastern part of the Baltic Basin. With farther retreat of the ice front, the North Sea flooded in across southern Sweden to produce the "Yoldia Sea," named for fossil molluscs found in the beach deposits that mark its former shores. As the ice continued to shrink, tilting and warping of the earth's crust caused the salt water of the Yoldia Sea to be replaced by the fresh water of "Ancylus Lake," also named for a fossil, and it in turn gave way to the "Littorina Sea" which finally became the Baltic Sea with its northward extension in the Gulf of Bothnia.

Almost everywhere the shorelines of these bodies of water, like those of the former lakes in the North American Great Lakes region, are marked by wave-cut cliffs, beach ridges, and deltas. At one place in southern Finland is an extraordinary shoreline feature. It is known as the "Jatinkatu"—the "Devil's Highway." Standing 390 feet above sea level, it was formerly the shore of the Baltic Ice Lake. The powerful waves and currents of that lake carried away all the clay, sand and pebbles from the moraine along its shore and concentrated the boulders and cobblestones in this unusual beach ridge.

Convict Lake in the Sierra Nevada, California, hemmed in by lateral and terminal moraines deposited during the Great Ice Age, is a glacial lake. (William A. Garnett)

ICE AGES OF THE REMOTE PAST

Extensive glaciation, like that of Pleistocene time, has occurred at widely separated intervals during the long history of the earth. The record of those ancient refrigerations consists of masses of *tillite* (consolidated till) in the midst of the sedimentary or metamorphic rocks. Tillite is distinguished from other conglomerates by its matrix of clay or silt and its glacially striated cobblestones and boulders. In some places it rests on a glaciated rock floor or pavement; sometimes it is associated with fine-grained sedimentary rocks having varve-like laminations. The age of a tillite is indicated by its relation to other rocks whose age has been determined either by their fossils, by the

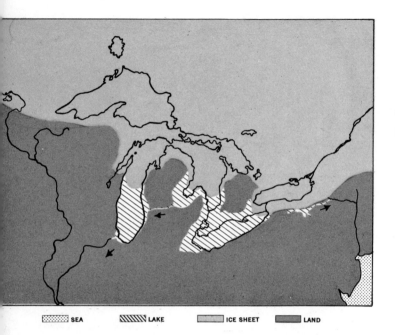
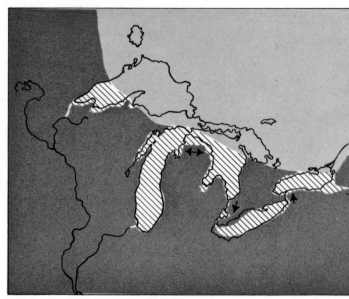

SEA   LAKE   ICE SHEET   LAND

Stages in the late glacial history of the Great Lakes of North America. About 15,000 years ago (above) an ice sheet covered the St. Lawrence Valley while the Great Lakes discharged down the Illinois River to the Mississippi. By 12,000 years ago (right) the ice front had withdrawn far enough to permit Lake Iroquois to discharge down the Mohawk River to the Hudson.

radioactive timekeepers, or by correlation with still other rocks whose age is known.

The oldest known ice age occurred far back in early Precambrian time, about two billion years ago. Tillites in "Eocambrian" or Archeozoic rocks are visible at several places in Norway, Sweden and Quebec; at the head of the Varangerfjord in northeastern Norway, the tillite lies on a glaciated pavement. There was also a glacial epoch about six hundred million years ago, indicated by tillites of latest Precambrian or earliest Cambrian age in China and elsewhere. But what was probably the earth's greatest ice age occurred in Early Permian time, about 270 million years prior to the Pleistocene's so-called Great Ice Age.

The distribution of the Dwyka tillite, Early Permian in age, indicates that most of Africa south of latitude 23° S. was covered with ice at that time. The direction of the striations and ice-carved grooves on the glaciated pavement, exposed at several places beneath the tillite, show that the ice moved generally south from the Transvaal. There is no certainty, however, that the African continent was then in the same position relative to Pole and Equator that it is today. Glacial deposits of similar age are also widespread in western and southern Australia and in Tasmania. An ice sheet stretching six hundred miles from east to west and a thousand miles from north to south is indicated by the Talchir tillite in the Salt Range and Rewah Province in India. Late Carboniferous or Early Permian tillites have been found at many places in Uruguay and southern Brazil and in the eastern ranges of the Andes in northern Argentina and Bolivia. One such outcrop in Bolivia is now only eighteen degrees south of the equator, well within the present tropics. Possibly the most interesting of all, a glacial tillite of this same general age has recently been discovered in one of the mountain ranges rising above the surface of the Antarctic Ice Sheet. Almost certainly the Permian ice sheets covered more of the earth's surface than did the ice sheets of the Pleistocene. Moreover, they spread much closer to the equator, *if* the continents were then located approximately as they are today.

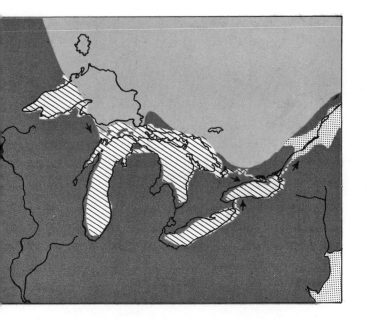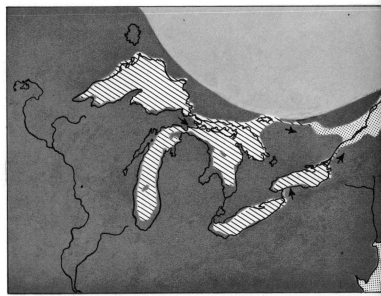

CAUSES OF GLACIAL CLIMATES

Many theories have been advanced to explain the widespread glaciations of the past. Most of them must be discarded; others still appear plausible, but none is wholly satisfactory. What caused the glacial climates is still a mystery.

The facts to be explained are these: at widely separated intervals in the long history of the earth, its usually equable and temperate climate gave way for a relatively short time to conditions that caused continental ice sheets to spread over certain parts of its surface far from the poles. During the Pleistocene and Early Permian epochs, if not also during earlier glacial epochs, each ice age involved sharp fluctuations between glacial and interglacial stages. These fluctuations were worldwide, which rules out theories attributing glacial climates to shifts of the earth's crust or of individual continents. They also indicate that changes in ocean currents or in the circulation of the atmosphere cannot be primary causes of such climates although they may influence the location and fluctuation of the ice sheets.

Three theories are still plausible. One is that the carbon dioxide, water vapor and dust in the earth's atmosphere create a kind of "greenhouse effect." Like the glass of a greenhouse, they retain the heat from solar radiation close to the earth's surface. If those constituents of the atmosphere were reduced in amount, the climate would be cooled, and vice versa. There is no doubt about their influence on climates, but there is skepticism about how effective it is.

The two remaining theories involve extraterrestrial causes and are weak because they rest on unknown factors. They are based on the fact that the climate of the earth is determined by the amount of energy it receives from the sun. The sun is a slightly variable star; if its energy output were reduced considerably below the minimum observed since man began to measure it, the result could be a glacial climate. This theory—fluctuation in the amount of energy emitted by the sun—is widely entertained by climatologists and glaciologists. There is at present no way to prove it or disprove it.

About 10,000 years ago (left) the retreat of ice in the St. Lawrence Valley allowed the upper Great Lakes to bypass Lake Erie and discharge into Lake Ontario. By about 9500 years ago (above) the upper Great Lakes were discharging directly into the St. Lawrence Gulf. Since that time, postglacial uptilting of the northeast corner of this area has left it as it is today.

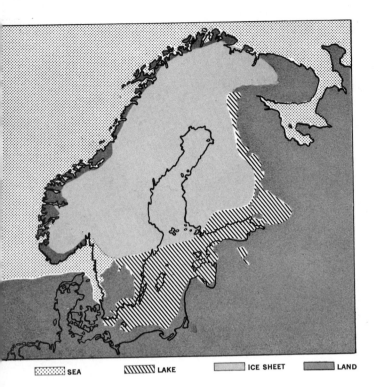

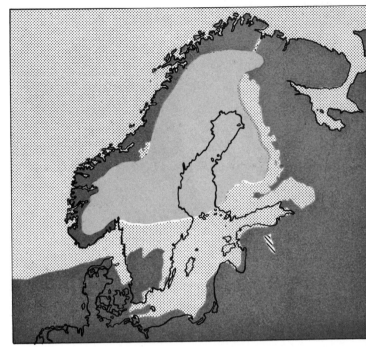

SEA   LAKE   ICE SHEET   LAND

Late glacial and postglacial stages of the Baltic Region. Map above shows the Fennoscandian Ice Sheet about 11,000 years ago when it covered most of Scandinavia and in the south impounded fresh water in the Baltic Ice Lake. The next map (right) shows a further shrinkage of the ice sheet about 9000 years ago that permitted the Ice Lake to drain off and give way to the Yoldia Sea.

The third theory stresses the fact that the amount of solar energy received by the earth is influenced by the particles of matter in the space between sun and earth. If earth and sun should move into a fairly dense cloud of cosmic dust, much of the sun's energy would be screened out before it reached the earth. Unfortunately for this theory, too, there is no way to test it. Astronomers cannot tell us whether we have been moving from a region of considerable cosmic dust to one less dusty during the last twenty thousand years. Nor can they tell us whether the earth has passed through four clouds of cosmic dust of appropriate density and dimensions in the last million years. The cosmic dust cloud theory is thus even more speculative than the theory of fluctuation in the sun's output of energy.

Even so, it is worth thinking for a moment about what future changes in the earth's climate would mean for mankind. If the trend of the past hundred years continues for the next few centuries, we will regain the weather of the Climatic Optimum of five thousand years ago, when mean annual temperatures in middle north latitudes were as much as 2 degrees Centigrade (3.6 degrees Fahrenheit) above those prevailing today. Sea level, which has risen about three inches since 1850, will rise another six or eight inches, but that will not be too great a threat to coastal dwellers. If thereafter the world climate returns to that of the Sangamon interglacial stage, as it may do in another five to ten thousand years, it will be a mixed blessing. Much if not all of the Antarctic and Greenland ice sheets will melt away and millions of square miles of icebound or otherwise inhospitable land will become available for settlement and exploitation; but sea level will be raised a hundred feet or so the world around by the return of meltwater, and many coastal cities and farms will be submerged. (It is an open question whether the inhabitants of New York and London in A. D. 10,000 will build lofty dikes to enable them to remain where they are, or move *en masse* to higher ground.

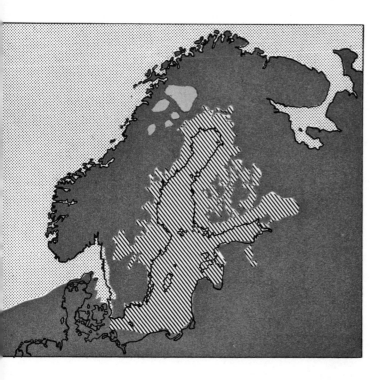 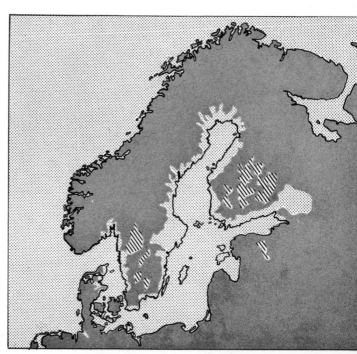

I think the odds are slightly in favor of such a rise in world temperature during the next several thousand years. Beyond that, no one should place any bets. The earth's climate may continue to warm slightly for hundreds of thousands of years, returning to the genial warmth of Tertiary time, one million to seventy million years ago, when the mean annual temperature in middle latitudes was some 10 degrees Centigrade (18 degrees Fahrenheit) higher than it is now. Or it may be that another glacial stage is in the offing, even as the Wisconsin glacial stage succeeded the Sangamon interglacial. If so, the climate will begin to cool some ten to fifty thousand years from now, and ice sheets will again spread over northwestern Europe and creep southward from Canada into the United States. The ice will advance at a rate of about 150 feet per year, giving plenty of time for migration; the loss of living space will be somewhat compensated by emergence of the now shallowly submerged continental borders. Our ancestors, equipped only with crude stone tools, weathered the Würm glacial stage in Europe and Asia; surely our descendants will be able to adjust their lives to a future ice age, if it comes.

The third map (left) shows the last remnants of the ice sheet about 7500 years ago, and the crustal warping that dammed the mouth of the Baltic Gulf and impounded the fresh water of Ancylus Lake. The fourth map (above) shows the crustal warping about 6000 years ago that resulted in the Littorina Sea, predecessor of the present Baltic Sea and Gulf of Bothnia.

10

Going Underground

Caves must be explored even as mountains must be climbed, simply because they are there. The lure of the unknown is added to the challenge of difficulty and danger. Throughout human history caves have evoked an emotional response, repelling many because of the darkness and the frightening feeling of being confined, but fascinating others by the mystery of dark passageways, by the beauty of the mineral deposits, and occasionally by their bizarre formations or eerie grandeur. Legends and superstitions concerning caves are found in many places; they were the abode of sybils and nymphs in Roman mythology, the entrances to the spirit world of eternal punishment or happiness in primitive religions, the places where the heroes of the Arthurian traditions sleep their long sleep.

Few of the so-called "cave men" of paleolithic times actually lived in caves; instead they inhabited shallow recesses below ledges of overhanging cliffs or only the entrances to deeper caverns. They penetrated into the interior passageways primarily for religious purposes or for rituals connected with magic. The modern speleologist, as the student of caves is called, is sometimes an archeologist specializing in the study of the prehistoric arts and customs recorded in those hidden places. Far more numerous are the adventurers who make a hobby of exploring caves and take pride in the less erudite designation of "spelunkers." Indeed, "spelunking" is today almost as popular as mountaineering. Using electric lights, access stairs, trained guides and even elevators, the exploitation of caves has now become a commercial enterprise of no mean proportions.

Beyond the urge to explore subterranean passageways and find out how they were formed, there is another good reason why we should go underground, so to speak, in our endeavor to become better acquainted with the earth. Most of the water used for drinking and cooking by the great majority of people throughout the world comes either directly or indirectly from beneath the earth's surface; and an adequate water supply is rapidly becoming the limiting factor in population growth in many localities. People living in rural surroundings are usually well aware that their potable water comes from wells and springs; people in urban areas know that somewhere, not too far away, there are reservoirs connected by aqueducts and pipes to the faucets in their homes. Although the usual reservoir receives some water directly from rain and surface runoff, much of its supply comes from springs and seepage from underground. Beyond its significance in meeting one of man's fundamental biologic needs, underground water—or more simply, *ground water*—is an important geologic agent, responsible not only for caverns in the earth but also for the forms of the landscape in certain localities and for such phenomena as geysers and hot springs.

GROUND WATER

The importance of the water occupying the cracks and pores in the outer part of the earth's crust is indicated in the large volume of water in springs, seepages and wells on every continent. Although the ground may be dry for a few feet below the surface, it is common in quarries, mines, and well-drilling operations to find rocks thoroughly saturated with water at shallow depths, usually less than a hundred feet and almost always less than a few hundred feet. Most of the water which fills the interstices of the rocks in this *zone of saturation* has trickled downward from the surface where it came from rain or melting snow and ice.

The upper limit of the zone of saturation is the *water table*. This has an irregular surface that tends to conform to the land surface of the region. It is higher beneath hills than beneath lowlands but it is usually much nearer the surface of the ground in the lowlands. It coincides with the earth's surface at the seashore, in swampy areas, and along the banks of permanent streams. It is of course higher than the surface of the ground wherever there are permanent ponds or lakes. Water tables rise and fall with changes in the weather. Ordinarily the water table is higher during wet weather than in long dry periods. It is customary, therefore, to use the term to indicate the average position of the top of the zone of saturation throughout a considerable period of time, in much the same way as the term sea level is used to indicate the mean level of the sea during many years, regardless of tides and winds.

The pores, crevices, and other open spaces in the ground above the water table are only occasionally filled with water. Much of the time they are occupied by air, and this part of the earth's crust is therefore known as the *zone of aëration*. Much of the water in this zone is seeping, trickling, or flowing downward from the surface toward the water table. Some of it may be stationary, held in capillary spaces for a time, or present as films of moisture on the surfaces of granular particles or the sides of crevices. Some of it may be moving upward, drawn back toward the surface by evaporation or by the roots of plants. The moist air and oxygen-rich water in the zone of aëration are among the most effective agents of weathering, especially because ground water is never chemically pure.

Four different forces can cause movement of ground water. In the order of their probable effectiveness, gravity comes first, then capillary action, gas expansion, and rock compaction.

Capillary action takes place only in the tiniest fissures or tubes. It tends to draw water into smaller openings, whether upward, downward or sideward. It cannot cause flow through such openings unless it is assisted by evaporation or by the roots of plants. As a result of capillary action, water tables are almost always topped by a fringe of moist ground extending several inches above the table. Consequently it is not necessary for the roots of plants to reach the table itself in order to secure moisture. If the roots

Above: In Mammoth Cave, Kentucky, stalactites dripping from the ceiling have joined stalagmites building up from the floor to form columns. (W. Ray Scott: National Park Concessions)

Above, right: Stalactites, stalagmites and dripstone drapery conceal the original limestone walls and roof of the Queen's Chamber in Carlsbad Caverns, New Mexico. (Josef Muench)

reach down to the *capillary fringe,* the plants can survive, even when the soil on the surface becomes dry during a period of drought. That completely explains why the vegetation of deserts generally has astonishingly long roots.

Gravity causes water to "seek its own level" beneath the surface of the ground precisely as it does on the surface. Hence, movement under gravity is primarily downward or sideward. Under special conditions, such as in artesian wells, ground water may develop hydrostatic head and flow upward, still "seeking its level" in response to gravity, just as the fluid does in a teapot, the spout of which fills from below.

Except where there are long tubular channels, such as in subterranean caverns, the flow of ground water is greatly retarded by friction between the percolating drops or trickles and the walls of the tiny passageways through which they must make their way. This explains why the water table is not horizontal like the surface of a lake, even months after the last rain has drenched a hill. In rocks of low permeability and especially in the deeper portions of the zone of saturation, ground water is nearly or completely stagnant. Highly permeable rock bodies, such as sandstone composed of large, even-sized grains or layers of unconsolidated sand or gravel, are known as *aquifers,* because they permit fairly rapid movement of water through their interstices. Comparatively impermeable rock bodies are known as *ground-water dams.* Impermeable rocks generally have low porosity, but some highly porous rocks have low permeability because their pores are not interconnected. Ground-water dams deflect the circulation of ground water and are important in determining the location of springs and wells.

Another factor in the movement of water underground is gas expansion. This tends to push water from spaces in which the gas is generated toward others which are less confining. An upward movement frequently results. The gas may be steam produced by the water itself at high temperature or it may be generated by chemical reactions such as those occurring in the transformation of buried organic matter to petroleum and "natural gas." Expansion of steam is the primary force behind the eruption of geysers and has much to do with the bubbling of hot springs.

The compacting of rock bodies also tends to squeeze liquids out of the spaces between the solid particles and causes movement that is more likely to be upward than in any other direction. When wet clay or silt is compacted to form a shale or siltstone, the result is a highly impermeable rock that generally behaves like a ground-water dam. It is still saturated with water but that water is firmly held within its minute pores and crevices by capillary action. A layer of sand is transformed into solid sandstone by a cementing process, with little or no compaction. To compact it and squeeze water from its pores would require enough pressure to crush the grains of sand or deform them by solid flow. Such pressure requires the weight of an overburden several miles thick or its equivalent in the horizontal movement of mountain-building.

SPRINGS AND WELLS

In a very real sense, wells are artificial springs. The relation of both to the local water table is the same. Most springs are in depressions where the ground surface intersects the water table and generally they are at places where the ground is more permeable than elsewhere. Thus, springs are most frequent at the foot of hills and along valley bottoms. Wet-weather springs are more common than those that supply water the year round. The water table flattens out or becomes lower as the dry season advances and it may no longer intersect the surface as before. Many an oasis in the deserts of Asia Minor and north Africa is fed by wet-weather *depression springs*. Digging a pit to reach the lowered water table in the dry season is an obvious procedure; many of the wells referred to in the Old Testament and other ancient records were the result of that simple application of common sense.

The great majority of the water wells on farms and in rural areas of such humid regions as eastern North America and western Europe are of this kind. A hole is dug or bored to penetrate the subterranean water table; water enters the hole under the direct action of gravity, seeking to maintain the level of the water table in spite of withdrawals from the well. Rates of flow are determined by the permeability of the rock bodies through which the well penetrates. Because of the great variability between such unconsolidated materials as clay, sand and gravel, or boulder-clay till and sandy till, and between shale and sandstone, some of these wells receive more abundant inflow than others. There is a widespread misconception that wells "tap an underground stream," but in fact the inflow comes from seepage or percolation from the zone of saturation beneath the water table. To secure an abundant, year-round supply, wells must be sunk to depths of several feet below the dry-weather position of the water table. Typically, such wells are on valley floors or in swales between hills, where the water table naturally comes closest to the land surface.

Springs and wells may also be located high on valley slopes or far up on a hillside, especially in regions of flat-lying sedimentary rocks. The great variation in permeability of beds of shale and of sandstone is often the controlling factor. Impervious shales act as ground-water dams and the water in overlying permeable sandy beds moves sideways instead of down, emerging on hillsides just above the outcrop of the impermeable layers. Because such springs are at the contact between rock formations of notably different permeability, they are called *contact springs*. Wells supplied by water in a similar manner are known as *contact wells*. To take maximum advantage of such conditions, they should be located on slopes above the line of natural seepage or wet-weather springs.

Movement of ground water in limestone may be quite different from that in other rock bodies. Water that has picked up organic acids from decomposed plants can easily widen the joint planes and bedding planes of limestone into tubes through which it may flow in veritable streams. Thence it may emerge at the surface in a *tubular spring* such as Silver Springs in Florida. With its average flow of seven hundred million gallons of water per day this is probably the largest single spring in the United States. Here the water issues from limestone into a crystal clear pool and then flows off to the sea in a sizable river. La Fontaine de Vaucluse near Avignon in France, with a similarly copious flow, issues from the foot of a limestone cliff and is the source of the Sorgue River. Countless other but smaller tubular springs occur in limestone regions such as the Karst in Yugoslavia, the Causses in south central France, and the Shenandoah Valley in the United States.

Wells dug or drilled in limestone are likely to be *tubular wells*. Their water level fluctuates greatly with variations in local rainfall, and the water is "hard" because it contains much dissolved calcium carbonate. Tubular wells should be kept under constant surveillance; on more than one occasion typhoid epidemics have been traced directly to them. Ground water may move directly from polluted soil into tubular openings and then into such wells, without the beneficial filtration to which water seeping through sandy aquifers is subjected. Any spring or well whose fluctuations in flow or level or water temperature synchronize fairly closely with changes in the local weather should be analyzed by competent authorities.

Still another kind of spring and well is recognized by students of ground water. For example, thick beds of sandstone crop out in the foothills of the Rocky Mountains and then dip down to the east and flatten out beneath the plains. They are excellent aquifers and are overlain by impermeable shale that serves as a ground-water dam. The aquifer sandstone beds are saturated with water that enters them near the mountains at altitudes of four thousand to eight thousand feet. At lower elevations the water presses upward against the confining beds of the overlying "dam." In more technical terms, it develops a hydrostatic head. If a well, drilled at altitudes of less than four thousand feet in these plains, penetrates through the shale into the sandstone, the water will rise under hydrostatic pressure. Although it tends to rise to the level of the intake near the mountains, it will fall short of that level because of the loss of pressure by friction as it moves through the aquifer. Such a well is an *artesian well;* springs that move along natural passage-ways through confining beds are *artesian springs.* (The name comes from the Artois region of France, famous long ago for its many flowing wells.) Technically an artesian well is any well, regardless of its depth, in which the water rises to a greater altitude than that of the local water table, whether or not the water overflows at the surface.

Both London and Paris happen to be located in artesian basins. The permeable layers of chalk that form the Chiltern Hills dip downward to the south in the Thames Valley where they are overlain by impermeable layers of London clay and then bend upward to reappear in the North Downs, along the southern rim of the London basin. The chalk aquifer, sealed in by the London clay, is at depths of only a few hundred feet beneath the city and some of the earlier wells drilled down to it were flowing wells. Its water content and replenishment from the Chiltern Hills and North Downs areas are, however, far from adequate to meet the present needs of London, which now derives much of its water supply from distant reservoirs. The Paris basin is much more extensive. The artesian aquifer is a poorly cemented sandstone known as the Greensand and the overlying confining bed is the Gault clay, a shaly formation just compact enough to serve as an effective ground-water dam. The Greensand is 1800 feet below Paris but rises gently to appear at the surface in the Argonne, 125 miles and more to the east, at altitudes of about 400 feet above the sea. There is thus sufficient hydrostatic head so that water gushes into the air from wells drilled down to this aquifer in Paris. Even so, the artesian water supply must be amplified by surface reservoirs and river water to meet the needs of that great city.

It is now known that artesian aquifers lie beneath some parts of the Sahara and the deserts in Arabia. Indeed it is highly probable that a few of the oases in those deserts are actually nourished by artesian springs. With further knowledge of subsurface geology in those regions, now accumulating from geophysical studies in connection with petroleum exploration, it is possible that in the near future the exploitation of artesian water resources will change the local economy as drastically as does the exploitation of petroleum resources.

CAVES AND CAVE DEPOSITS

Most caves resulting from the work of ground water are in calcareous sedimentary rock (limestone or dolomite). They formed when water circulating along the bedding planes and joints in generally flat-lying strata dissolved some of the rocks. In many caves the network of narrow corridors is obviously the result of enlargement by chemical solution along the original joint planes until they have become open passageways. Some caves have two or more levels of such passages, each corresponding to an original bedding plane where water circulated with relative ease. Relations between the pattern of the "rooms" in large caverns and the original pattern of the parting planes are often difficult to detect but they generally are clear in the smaller corridors. The only major problem for anyone who wants to find out how caves came to be is their relation to the water table at the time of their origin.

All caves that can be entered are in the zone of aëration above the water table. Where there is a subterranean pool or "River Styx" in the lower part of a cave, the surface of the open water reveals precisely the position of the water table at that place. Occasionally "spelunkers" have used skin-diving equipment to reach recesses not otherwise accessible. But the generalization holds: caves are in uplands where the top of the zone of saturation is far enough below the surface of the ground to leave room for them. There is, however, little or no reason to think that this was the situation at the time the caves were formed. Indeed, most investigators agree that as a rule caves are largely a result of solution below the water table rather than above it.

According to this view, the history of the usual cave began long before the local landscape had attained its present form. Then the water table was much nearer the surface than now. Slow circulation of water in the zone of satura-tion enlarged partings along bedding planes and joints in the limestone or dolomite. There was little or no mechanical erosion. Instead, the water that filled the tubular openings attacked and dissolved ceilings and walls as well as floors. The insoluble silty particles released from the soluble rock dropped through the slowly moving water to the floor of the cavity, producing the smooth pathway found in some corridors today. After the rooms and pas-sageways had been dissolved out, the adjacent surface streams deepened their valleys, lowering the water table and draining the cave, thus permitting the air to enter it for the first time. Enlargement of the cave ceased and the water seeping downward from the surface has built dripstone deposits such as stalactites and stalagmites that partially fill many caves and give them some of their most fascinating features.

To be sure, in some caves underground streams have eroded the rock during the second phase of their development after the water table has dropped, but the features produced by such erosion are completely in con-trast to the far more widely distributed effects of solution. Actually the second phase is one of cave deterioration, although one hesitates to apply that term to the often beautiful and fantastic deposits that are gradually filling up many famous caves. It can be applied without apology, however, to the fall of ceiling rock and the debris washed into caves by wet-weather streams.

Collapse of part of a cave roof has its brighter side. Horizontal entrances to caves at floor level are often blocked by debris that slides down exterior slopes, and it may be easier to enter them by holes in the ceiling. Thus, the entrance of the Gouffre de Padirac, one of the most famous caves in south central France, is through a roughly circular hole, 96 feet in diameter at the top but much larger below. The visitor descends 125 feet in an elevator to the debris left by the collapse of the roof and then down irregular steps for another hundred feet. This brings him to a gently inclined corridor along which he walks until he is about 300 feet underground. Here he reaches the present water table; a slowly moving stream fills the gallery from side to side. He floats down this river in a small boat between steep walls that disappear in the darkness overhead until a quarter mile downstream he reaches the Lake of the Raindrops, so-called because the walls sparkle with tiny crystals of calcite. Here he sees the great dome, 180 feet long, 75 feet wide, and 280 feet high, the largest of the chambers in this extensive under-ground system of rooms and corridors.

69. The Grand Palace in Lehman Cave, Nevada. The stalactite hanging at right almost meets the stalagmite building up from the floor. The massive stalagmite left of center displays the "pulpits" sometimes formed by dripping water saturated with mineral matter. (Josef Muench)

192

70. Carlsbad Caverns (left), New Mexico, are sumptuously decorated with dripstone formations of both stalactites and stalagmites. (Josef Muench)

71. Hot springs issuing on a hillside usually form terraces like those in Opal Spring Terrace (above, right), Yellowstone National Park, Wyoming. Each successive pool of water develops a rim of precipitated mineral matter, often colored by iron oxides and the algae that thrive in the warm water. (Josef Muench)

72. Beatus Cave (right), near Interlaken, Switzerland, is adorned with dripstone draperies whose intricate pattern indicates the routes of driblets of water depositing mineral matter. (Werner Luthy)

73. The tree trunks and this natural bridge in the Petrified Forest National Monument, Arizona (above), have been so completely silicified by ground water that they remain even when the clay in which they have long been embedded is washed away. (Josef Muench)

74. The famous Natural Bridge (left) in Virginia is all that remains of the roof of a narrow cavern dissolved out by ground water in the limestone beds of this part of the Shenandoah Valley. (Benjamin M. Shaub)

75. The throat of this geyser (right) in Yellowstone National Park, Wyoming, is surrounded by mineral matter deposited when spatters of steaming hot water cooled and evaporated. (Robert Clemenz)

76. The banks of Firehole River (below) in Yellowstone National Park, Wyoming, are covered with brilliantly colored mineral matter deposited by hot water from geysers and springs that cooled and evaporated while trickling over the ground. (Robert Clemenz)

77. Some of the Hot Springs Terraces near Thermopolis, Wyoming (right, above), form gigantic bowls when the mineral matter dissolved in the hot water reaches the cooling edges of the pools and builds up the rims. (James R. Simon)

78. Similarly the Punch Bowl Spring in Yellowstone National Park, Wyoming (right, below), owes its beauty to the mineral-laden water that flows continually from vents within the bowl. (Josef Muench)

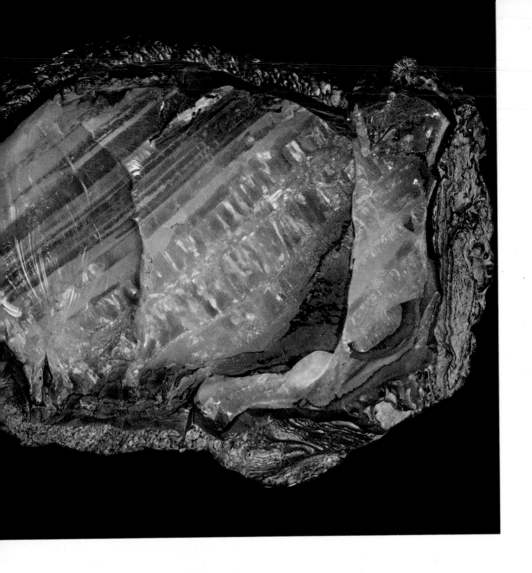

79. Fragment from a vein of gem-quality banded opal (left), popularly called Harlequin opal, from Australia.

80. Malachite (center) from the Ural Mountains, Soviet Union. It is valuable as a copper ore mineral and even more as a gem stone and in ornamental mosaics.

81. Granular aggregate (below, left) from a mine near Superior, Arizona, consists of bornite (a copper ore mineral popularly called "peacock ore"), chalcopyrite (another copper ore mineral), and milk-white quartz.

82. A cut and polished agate (below) consisting of concentric bands of vari-colored chalcedony filling a cavity in an igneous rock from Uruguay.
(All by Benjamin M. Shaub)

Certain of the chambers in Mammoth Cave, Kentucky, are nearly as large as those at Padirac, and some in Carlsbad Caverns, New Mexico, are even larger. There are dripstone deposits in all three, but the stalactites, stalagmites, columns and draperies are more elaborate in Carlsbad than in the others. Such ornamentations are also found in countless smaller caves throughout the world, many of which are famous for the beauty and variety of their decorations.

Dripstone deposits, in all their limitless variety of form, are composed almost exclusively of tiny crystals of the calcium carbonate mineral, calcite, that makes up the bulk of all limestones. It is readily dissolved by water containing carbonic acids or carbon dioxide, picked up as it trickles past plant roots and through soil. Soon after it enters crevices in limestone, the water becomes saturated with calcareous mineral matter in solution. Seeping downward through the roof of a cave, the ground water emerges along cracks in trickles or as individual drops. Exposed to the air in the cave, it tends to evaporate or lose its carbon dioxide, causing it to deposit tiny crystals of calcite. Thus stalactites grow downward from the ceiling like icicles, and draperies hang pendant where a long crack concentrated the seepage in a narrow zone. The water that drops from the tip of a stalactite builds a stalagmite upward from the floor below. Sometimes the two are joined to make a column. The diversity of form of these dripstone features is as great as the ingenuity displayed by the imaginative names that have been given them in many caves (Plates 69, 70 and 72).

A few caves are so situated that water freezes in them to produce forms of ice just like the calcareous formations. Such caves generally occur in snow-capped mountains: the Eisriesenwelt in the Tennengebirge, not far from Salzburg, and the Rieseneishöhle, near Obertraun at the foot of the Dachstein Massif in Austria, are excellent examples. It is no stretch of the imagination to describe a stalactite as icicle-like; the two are formed in identical fashion. Instead of evaporating to leave behind tiny crystals of calcite, the water seeping through the roof of an "ice cave" congeals into tiny crystals of ice. The important factor is temperature. Unless there is rapid circulation of air through the passageways of a cave, the temperature will at all times closely approximate the mean annual temperature at the surface. Thus in winter the air in a cave is warmer than the surface air, whereas in summer it is colder. The year-round temperature in a cave at or above snow line is likely to be slightly below 32 degrees Fahrenheit (0 degree Centigrade). Summer melting of the surface snow liberates water at a temperature just above freezing. When that water seeps through the cave roof and comes in contact with the slightly colder air in the cave, it promptly freezes; the result is "dripstone" composed of what is truly a mineral: crystallized H_2O.

KARST TOPOGRAPHY

Regions under which lie extensive beds of limestone or dolomite generally display a topography which is quite different from the ordinary stream-carved landscape. In such regions, most of the surface water descends quickly into underground passageways or seeps downward and drips from the ceilings of caverns. Surface streams—except possibly an occasional river in a

A "gypsum rose" in Mammoth Cave, Kentucky, was formed by ground water saturated with calcium sulfate. (W. Ray Scott: National Park Concessions)

deep valley—are few and short; ill-defined valleys end abruptly in sinkholes. The features of the landscape are mainly the result of solution by ground water rather than of erosion by running water. The name *karst,* applied to such land surfaces, comes from the region northeast of the Adriatic where such features are common.

The sinkholes in a region of karst topography are of two types. Some are saucer-shaped depressions known as *dolines* which are hundreds of feet wide, slope gently inward, and are often covered with soil and vegetation. They generally have no entrance to underground passageways. Instead, their bottom may be so covered with fine silt or clay as to maintain a pond in the rainy season or even throughout the year. On occasion, such a pond may abruptly lose all its water in one downgulp. This type of sinkhole is the result of downward solution from the original surface.

The other type of sinkhole is steep-walled, commonly discloses bare rock, and may open downward into an extensive cavern. It is formed by the collapse of the thinned roof of a cave beneath it. Geologists call this a *ponor,* or use the more picturesque term from the south of France: *gouffre.* The collapse of the roof of a long narrow cave may leave a remnant standing as a bridge across a narrow, steep-walled valley. This is the origin of the famous Natural Bridge in Virginia (Plate 74).

HOT SPRINGS

Although the water of most springs has a temperature close to the mean annual temperature of the air at the particular locality, at many places on every continent the springs are hot, and sometimes close to the boiling point. Hot water is much more active chemically than cold water; hence almost all hot springs discharge water containing much mineral matter in solution. The dissolved salts—mainly carbonates, bicarbonates, sulfates, sulfides, and chlorides of calcium, sodium, iron and magnesium—are generally noticeable and sometimes unpleasant to the taste. If sulfates and sulfides are abundant, the water will probably smell like rotten eggs. Mineral springs, especially if warm or hot, have real or fancied medicinal value and many, such as those at Vizela, Portugal, Aix-les-Bains, France, Baden, Switzerland, Wiesbaden, Germany, and Hot Springs, Arkansas, have long been famous health resorts.

Heat for ground water comes from several sources. The temperature in mines, tunnels and wells always increases downward below a depth of forty or fifty feet. If the window of the train compartment is open when going through the Simplon Tunnel in the Alps, or any other long mountain tunnel, one notices warmer air midway between entrance and exit, where the mountain peaks are thousands of feet directly above. The rate of temperature increase with depth below the surface—the thermal gradient within the earth's crust—differs greatly from place to place, but it is usually about one degree Fahrenheit per 65 feet (one degree Centigrade per 35 meters). Recalling the low velocity at which ground water ordinarily percolates in the zone of saturation, it is obvious that a hot spring must be fed by water that has either moved upward with extraordinary rapidity from depths of several thousand feet or has been in contact with extraordinarily hot rock at depths of only a few hundred feet. Both are possible; the latter is more probable.

The majority of hot springs are in localities where there has been volcanic activity in geologically recent time. Presumably such springs derive their heat from igneous rock, either extrusive or intrusive, that is at no great depth and has not yet completely cooled from its original consolidation temperature. Water seeping down from the surface until it is close to such rocks is heated by them and may return to the surface in aquifers or tubelike passageways before it has lost all of its heat. Mixed with that *meteoric water*—so-called because it was in the atmosphere before it entered the ground—there may be relatively small amounts of *juvenile water*, that is, water newly born within the earth. As molten magma crystallizes into igneous rock, the water it originally contained is squeezed out by the growth of the crystals. Such water adds not only heat but also its dissolved mineral matter to the meteoric water with which it mingles. That it has done so in a few hot springs is revealed by the presence of compounds of boron, arsenic, and other elements not found in rocks through which meteoric water had passed.

Another source of heat is found where the earth's crust has recently been subjected to the kind of crustal movements that are involved in the making of mountains. Any movement of one segment of the crust past another gives rise to friction, and friction generates heat. Rocks made hot by friction may continue for a long time to warm the water circulating through them. Still another possible source of heat is the chemical reactions that may be going on anywhere within the earth's crust. Indeed, ground water itself may produce heat by such reactions as oxidizing sulfides to sulfates. Finally it is possible that local concentrations of radioactive elements may raise the temperature of their surroundings and contribute to the warming of ground water. However, few if any so-called "radium springs" contain enough radioactive elements to make that explanation plausible.

Whatever the cause of their high temperature, the waters of hot springs are especially potent agents for bringing mineral matter to the surface and depositing it there. The deposits are the result of chemical and sometimes biochemical precipitation from solution, expedited by rapid cooling when the water comes into contact with the cooler air, by loss of contained gases when the water is no longer confined within the crannies of the rocks, by rapid evaporation if the air is dry, and by the life processes of algae and other organisms that may be present.

Broad, low domes of calcareous materials surround many hot springs; occasionally terraces rise steplike on hillsides (Plate 71). Pools of water are surrounded by low rims over which the water spills to deposit fluted draperies on each successive descent. In some places, as in Yellowstone National Park in Wyoming, the hot spring terraces are unusually large and spectacular (Plates 77 and 78).

GEYSERS

Hot springs display every gradation from a quiet flow or a pool that is troubled by escaping bubbles of steam and occasionally boils over, to a violent periodic eruption of a column of water and steam. The last is known as a *geyser,* a name of Icelandic origin.

Although in certain volcanic regions in Malaya, Japan, and South America there are boiling springs from which water frequently spurts upward, geysers worthy of the name are found only in three localities: Iceland, the province of Auckland in New Zealand, and Yellowstone National Park in the United States (Plate 75). The geysers in Iceland and New Zealand are not nearly so active today as they were a few decades ago, and those in Yellowstone are beginning to show signs of deterioration. Yellowstone's Old Faithful, for example, formerly sent up a column of water 125 to 150 feet high at intervals of sixty or sixty-five minutes, each eruption lasting about five minutes; at present its eruptions come at less regular intervals and frequently do not exceed a hundred feet in height.

Geysers are so rare because they require special subterranean conditions. The tubes through which the water rises must be fairly large and open but so shaped—whether crooked or branching or irregular in width—as to inhibit the convection currents that normally equalize temperatures throughout a column of water. The important requirements are unusually hot rock bodies at depths of several hundred feet, an abundant supply of ground water, and passageways that permit rapid flow under pressure from gas expansion but not from convection. The boiling point for water varies greatly with pressure—212 degrees F. (100 degrees C.) at sea level under normal atmospheric pressure; 291 degrees F. (144 degrees C.) under the weight of a column of water one hundred feet high; 394 degrees F. (201 degrees C.) under the weight of a column of water five hundred feet high; and so on. The familiar pressure cooker operates on this principle. Under appropriate conditions the water in a geyser tube may be so heated that its temperature from top to bottom is only a few degrees below the boiling point at every depth below the surface. If at some fairly deep place, the temperature rises to the boiling point, whatever it may be—400 degrees F. or 500 degrees F. or more—the water there will flash into steam with all its tremendous expansion force. The water above will be pushed suddenly upward, some overflowing at the

geyser vent, but much of it rising to places at which its temperature is equal to the boiling point. Thus a considerable part of the water will vaporize at approximately the same time and violent eruption will be inevitable. This cleans out the entire system; the geyser cannot erupt again until its tubes have been filled anew by infiltrating ground water and that water has attained the necessary temperature.

One or both of two changes cause geysers to be short-lived, geologically speaking. The hot rock cools fairly rapidly, much more swiftly by the circulation of water and steam than by conduction through overlying rocks, and constrictions in the geyser tubes are removed by the eroding action of the gushing water. In time the passageways open so wide that convection currents circulate freely enough to prevent explosive transformation of water into steam. The geyser becomes merely a boiling spring. One should not postpone a visit to Iceland, New Zealand or Yellowstone; in two or three hundred years they will probably display only hot springs, paint-pots and boiling pools.

In Yellowstone National Park, Wyoming, the ever-enlarging terraces formed by hot springs surround and kill many trees. (Tad Nichols: Western Ways Features)

CONCRETIONS, NODULES AND GEODES

The highly selective chemical action of ground water sometimes results in the formation of *concretions* in the midst of a sedimentary rock. These are of all sizes and differ greatly in mineral content from the surrounding rock.

205

Originally their ingredients were widely separated molecules of calcium carbonate in a shale, for example, or of iron oxide in a sandstone, or of iron carbonate in a limestone. Dissolved in slowly circulating ground water, the molecules precipitate from solution at one spot because of the presence of organic matter or other chemical factors. Small concretions tend to be enlarged by additions of the same substance, as "like attracts like."

Concretions composed mostly of silica are likely to be much more irregular in shape and are usually called *nodules*. They are often found in limestone and chalk formations. If the silica nodules are light brown, yellowish or light gray, they are *chert;* if dark gray or black, the substance is *flint*.

Occasionally a mass resembling a concretion, when broken apart, is found to be hollow, with crystals projecting inward, rather than solid with concentric layers. Such a roughly spheroidal hollow body, lined on the inside with crystals that point inward, is a *geode*. Geodes appear most frequently in limestone but may occur in shale. Generally more resistant to weathering than the rock in which they formed, they may be found lying loose in the soil or in surface debris mantling the solid rock. Some are eye-catching specimens, treasured by mineral collectors, offered for sale by curio dealers, or on display in museums.

The most common geodes are two or three to twelve inches in diameter and consist of a thin outer shell of chalcedony and a thicker inner shell of quartz crystals, many of them beautiful prisms terminating in hexagonal pyramids projecting toward the center. Calcite or dolomite crystals line the interior of some geodes and a variety of other minerals may occasionally be found in them.

Geodes originally formed in a cavity, such as the space within a fossil shell, from which the geode expanded. The layer of chalcedony is the hardened form of an original silica gel. Expansion was due to pressure from sea water trapped inside the clump of silica gel and happened when the calcareous oozes or muds were undergoing consolidation into limestone or shale. The crystals projecting inward were precipitated later from ground water that infiltrated the hardened, hollow spheroid.

PETRIFIED FORESTS

The many "fossil forests" in various parts of the world demonstrate the importance of underground water as an agent of geologic change. At the famous Petrified Forest in Arizona (Plate 73) there are hundreds of prostrate silicified tree trunks and stumps, exposed by the weathering away of the poorly consolidated dark brown or varicolored shale in which they were once buried. The majority of the trees are conifers, distantly related to certain "pines" now growing in South America and Australia. They are of Triassic age, roughly 180,000,000 years old. Associated fossil leaves and seeds indicate that they grew in a humid, subtropical climate along streams in a lowland savannah—an environment quite different from northern Arizona today.

It was once thought that in the petrifaction of woody tissues organic matter was replaced, molecule for molecule, by mineral matter dissolved in percolating ground water. It is now evident, however, that the molecules of silica actually fill the minute interstices within and between the walls of the plant cells. Petrifaction is really a process of mineral emplacement by which the network of woody tissues becomes impregnated with solid substance. Ground water accomplished the petrifaction in the zone of saturation before the wood had undergone much decay. The continuing presence of organic molecules in the petrified wood accounts for its varied and beautiful coloration. The large amount of silica, in the form of chalcedony, opal (Plate 79), and agate (Plate 82), suggests that hot ground water had dissolved the mineral matter from subjacent rocks and precipitated it from solution when rising toward the surface.

The source of the hot water responsible for the even more extensive petrified forests in northeastern Yellowstone National Park is obvious. Here the majority of the petrified tree trunks are standing upright, as they originally grew, but are enclosed in thick deposits of volcanic ash and cinders. Most extraordinary here is the succession of at least twenty buried forests, one above the other, in the midst of the layers of volcanic debris totalling more than two thousand feet in thickness. Evidently ground water percolating through the layers of ash and cinders soon after they were ejected from nearby volcanic vents was promptly heated and then gradually cooled. Like the water in geyser tubes, it would first dissolve and then precipitate silica; the tree trunks would be the most likely place for that precipitation.

The fossil forests in Yellowstone National Park are less than half as old as those in Arizona. They include more than a hundred different species of plants typical of a humid lowland environment in a warm temperate to subtropical clime—again something very different from the conditions in Yellowstone today.

METALLIC ORE DEPOSITS

Water in the ground has had much to do with the origin of the great majority of metallic ore deposits throughout the world. The emanations from crystallizing magmas that produced ore bodies in zones of contact metamorphism were partly gaseous and partly fluid. The fluid was largely juvenile water, but as it penetrated the crustal rocks above the magma chamber it mingled with the water that had percolated down from the surface and was filling the crevices and pores in the zone of saturation. The farther the juvenile water moved from its source in the solidifying magma, the more it was diluted by the meteoric water it encountered. Chemical precipitation of mineral matter from solution in such ascending, superheated liquids, with their varying proportions of juvenile and meteoric water, have been responsible for the formation of many valuable ore deposits. The ores at Cornwall, England, for example, were formed in this way; fairly close to the parent igneous rock are tin veins (now practically exhausted) that pass upward and outward into copper veins, followed by veins of lead and silver, then antimony, and finally iron and manganese carbonates. Similarly, the rich ores of zinc and lead in the Tristate District of Oklahoma, Kansas and Missouri consist of crystals of sphalerite (ZnS) and galena (PbS) deposited by ascending hot solutions in cavities in the limestones and dolomites fairly near the surface (Plates 80–81). And as a final example, of special interest today, the pitchblende (one of the most important ores of uranium) mined at

Shinkolobwe, Congo, is in veins deposited by similarly ascending hot fluids and therefore referred to as hydrothermal veins.

In these and many other ore deposits the metal-bearing minerals were so concentrated by the original hydrothermal solutions as to make it profitable to mine them. In some ore bodies, however, the original or "primary" deposit was not rich enough to cover the mining costs and these become valuable ores only if they are later improved by "secondary" enrichment. Here, ground water is the all-important agent. Moving downward through the zone of aëration and circulating variously in the zone of saturation, it often effects great changes in primary ore bodies. Sulfides of the metals are changed into soluble sulfates or carbonates; these are carried downward and redeposited below the water table, thus greatly enriching the ore body there. The process involves many complex chemical reactions but the net result is the conversion of many low-grade ore bodies into workable deposits.

In some instances the process of enrichment works in reverse. The high-grade iron ores of the Lake Superior District in the United States and Canada, for example, include certain ore bodies that have been "negatively" enriched by the action of ground water. Deposited originally in Precambrian time as sediments containing 10 or 20 per cent iron, they were later changed slowly into ores running 30 to 60 per cent iron, as ground water leached the silica from the rock, leaving behind the less soluble iron minerals, such as hematite (Fe_2O_3), in rich concentration.

Circulating ground water may also bring together in a few places metallic mineral matter originally distributed widely throughout a geologic formation. The most productive uranium ores in the United States were formed in this way. The ore mineral is the yellow, powdery carnotite, a hydrous vanadate of potassium and uranium. It occurs in the sandstones and sandy shales of the plateau country in Utah, New Mexico and Arizona, and is a secondary mineral resulting from the action of ground water on pre-existing uranium-bearing minerals. Relatively pure masses of it were concentrated near fragments of petrified wood or other vegetal matter because of the chemical reactions between the ground water carrying the uranium salts in solution and the organic compounds in those substances. Other workable deposits of carnotite are also being exploited in Katanga, Congo, and near Olary, Australia.

83. The mysterious "skating rocks" in Death Valley, California, were moved by strong gusts of wind across the mud flat when it was the bottom of a shallow pond, the surface of which was probably covered with ice. (Robert Clemenz)

84. A rockslide (left), in the Valley of Castle Creek, Utah, composed largely of fragments of sandstone that crashed down from the northeast side of Grand View Mountain. (Andreas Feininger)

85. A vast mass of water-soaked soil and rock fragments after plunging, in the form of stiff mud, down a steep-sided valley (above, right) in the Wrangell Mountains, Alaska. (Steve McCutcheon)

86. A rockslide (right) near Frank, Alberta, Canada, in 1903 left a huge scar on the side of Turtle Mountain, high above its heap of rubble. (William W. Bacon III: Rapho Guillumette)

II

Landslides, Avalanches and Mudflows

Many of the great catastrophes in human history have been caused by a sudden downward movement of materials lying on the surface of the earth. The avalanche that overwhelmed eight villages and killed more than thirty-five hundred people in Peru in January, 1962, was front-page news around the world. Seventy inhabitants of the coal-mining town of Frank in Alberta, Canada, lost their lives in 1903, when a landslide destroyed the greater part of that community. And the visitors to Elm in the Swiss Alps are still fascinated by accounts of the calamity that occurred near there in 1881: a large slate quarry had been opened near the base of the Plattenberghoff and evidently enough supporting rock was removed so that one afternoon great blocks broke loose from the east side of the mountain, crashed into the quarry, and spread in fragments across the gentle slope below. The debris almost reached an inn where sightseers were refreshing themselves. Seventeen minutes later a larger slide came hurtling down from the west side of the mountain and overwhelmed the inn and four houses adjacent to it. The sightseers and local inhabitants fled, but a score of them were killed. Four minutes after the second slide the whole mountain top suddenly fell directly into the quarry and then pitched forward almost horizontally across the valley. Some ten million cubic yards of rock crashed downward more than twelve hundred feet with such momentum that the shattered fragments covered a horizontal area of nearly three quarters of a square mile to a depth of thirty to sixty feet and climbed as much as three hundred feet up the opposite slope. Those who escaped the disaster reported that the fearful mass of rock "poured like a liquid" from the quarry site into the valley.

To such movements of earth materials *en masse,* in contrast to the movement of individual dust particles, sand grains, pebbles, boulders, and other rock fragments by running water, blowing winds, and sliding glaciers, geologists apply the designation *mass movements.* Some of these are almost imperceptibly slow, unlike the one just described. Some involve water-soaked ground rather than relatively dry rock, and some include more snow than anything else. Some take place on very gentle slopes, others only on steep hillsides. Although all may be considered landslides, it is better to restrict that term to the perceptible downward sliding of a relatively dry mass of earth, or rock, or a mixture of the two. Other mass movements may then be thought of as related phenomena to which such terms as soil creep, mudflow, and avalanche apply. But, as in all natural phenomena, there is an almost continuous gradation from one type to another.

213

The downslope creep of hillside soil is the most widespread, if least spectacular, manifestation of the direct effect of gravitation on land. It may occur on gentle slopes but is generally more apparent on fairly steep ones. In either situation it may proceed beneath a grass or forest cover. *Soil creep* is revealed by tilted trees, fence posts, or monuments, and by broken retaining walls.

Alternate freezing and thawing of water in the soil hastens soil creep. Frost heaves are familiar and troublesome demonstrations of the expansion power of freezing water, especially where they affect a highway or building foundation. When the disturbed materials on a slope settle back after the frost has melted, they always move downslope under the influence of gravity. Repeated frost action over the years can produce a significant migration of the soil.

In regions of permanently frozen ground, where "permafrost" (Plate 87) extends downward for tens of feet and only the upper foot or two melts each summer, frost-controlled creep often produces remarkable patterns in soil and rock debris. Most of the movement takes place soon after the thawing of the frost-riven ground in the spring, and the annual displacement of the creeping material is generally only a few inches. The pattern developed at the surface is a series of lobes or tonguelike masses of debris. Gullies, ravines, hollows, and even broad valleys may be completely filled with this *solifluction debris,* as it is called. Where solifluction (soil flow) affects a region in which the bedrock is mantled by glacial till or other unconsolidated mixtures of large and small rock fragments, the reticulated pattern of the surface may be especially prominent. The larger stones become segregated within the matrix of smaller pebbles, sand and clay, and appear in *stone stripes* or *stone polygons*. The stripes are more likely to develop on moderately steep slopes where they run downhill, the polygons on more gentle slopes. Both are common in polar and subpolar regions where permafrost is widespread. In formerly glaciated regions now free of permafrost, they are relics of climatic conditions near the dwindling ice sheets at the end of the Great Ice Age. Among the most striking are those in the so-called "alpine gardens" just below the summits of the highest mountains in the Presidential Range in New Hampshire.

MUDFLOWS

Mudflows constitute another special type of mass movement. "The Slumgullion" near Lake City, Colorado, is an impressive example. There, the "mud" is largely disintegrated volcanic rock. The entire upper half of the mountain ridge on the east side of the valley of Lake Fork of Gunnison River is composed of alternating lava flows, tuff, and agglomerate. In some places these firm rocks were notably altered by ascending gas and superheated water which were, at least in part, of magmatic origin. After the Lake Fork valley had been formed by stream erosion and sculptured by a glacier during the Great Ice Age, a huge mass of this altered rock was most precariously situated. Water from rain and melting snow promptly saturated its innumerable cracks and frost further disintegrated it. It was probably during a

thaw in late spring or early summer that the water-soaked mass "let go." It flowed like a viscous liquid from high on the mountainside, down a shallow tributary valley to the major valley where it spread laterally upstream and downstream. One of its many recurrent movements was rapid enough to splash some of the debris upward more than three hundred feet on the opposite side of the Lake Fork valley. The thick mud carried along many large blocks of rock; countless smaller fragments of greatly altered volcanic rock mingled with the pink and light brown mud. The major movements must have occurred several hundred years ago, but minor slumping continues to the present day at many places on the surface of the mudflow.

Because The Slumgullion moved clear across the main valley, it formed a dam, ponding the water above it and forming Lake San Cristobal. The lake's outlet has been deepened to bedrock where it crosses a low point in the surface of the mudflow, and the overflowing water cascades downward to reach the valley floor below the natural dam. Since that dam was formed, the rushing water has not had time to cut much of a channel in the bedrock. Thus, this mudflow is responsible for a rather lovely little waterfall and a very beautiful, mountain-rimmed lake.

The Slumgullion is extraordinary among mudflows for its large dimensions and the great changes it produced in the local topography, but its dynamic behavior is typical. Mudflows are always water-soaked and their movement is fairly rapid. The flow usually follows a drainageway on a slope, which may be quite gentle and is sometimes the path of recurrent mass movement.

Mudflows or earthflows have temporarily obstructed many highways in regions subjected to sudden heavy rains. It is not merely for esthetic reasons that highway engineers hasten to cover with vegetation the steep sides of road cuts in unconsolidated earth materials. Clay and glacial till are notoriously unstable when water-soaked, and thoroughly weathered rock of almost any type is nearly as bad.

A giant frost heave, called a pingo, rises three hundred feet above the Mackenzie River Delta in Alaska. Although its core is solid ice, its silt-covered flanks are mantled with lichens. (George Hunter: Ottawa)

215

The designation "quick clay"—analogous to quicksand—has long been applied in Norway and Sweden to clays that undergo abrupt change from a stable condition to turbulent motion. At many places in those countries extensive terraces, benches, and flat uplands of glaciomarine clay stand as much as six hundred feet above sea level as a result of postglacial uplift of the Scandinavian region. One of the most careful studies of movements of that kind of clay was made at Surte, Sweden, in 1950, when a "slide" twenty-two hundred feet long and twelve hundred feet across was formed in less than three minutes. The "quick clay" moved at a velocity of two to three miles per hour down a slope that was nearly everywhere less than one degree. Thirty-one houses, a railroad, and a highway were displaced. The formerly stable clay and silt, saturated with water from rains, probably started moving in response to concussions from a pile driver operating nearby. An almost negligible slope was adequate to cause the mud to flow. What had seemed to be an attractive residential area was suddenly transformed into a scene of desolation.

AVALANCHES

"Beware the avalanche" is a thought almost constantly in the mind of every mountaineer climbing toward snow-clad heights. In such regions as the Swiss Alps where the land is rugged and the snow abundant, it is a recurrent threat to everyone living in the valleys. Every year, in one or more of the earth's many lofty mountain ranges, catastrophic avalanches take their toll. In December 1916, for example, more than six thousand Austrian soldiers were killed in the Tyrol even though the danger was known in advance. And 1951 will long be remembered in Switzerland as an "avalanche year."

Many avalanches are "dry": they are essentially dense clouds of finely pulverized snow mixed with air. They may be started merely by gusts of wind swirling down from a ridge crest upon windrows of snow loosely banked on a fairly steep slope, or they may originate when a cornice of snow, precariously perched on the lee side of a windswept summit, suddenly crashes down snow-covered slopes. The mixture of air and snow particles has several times the specific gravity of clean air and consequently it may sweep down the mountainside with irresistible force at a velocity exceeding that of the wildest hurricane. Anything in the path of such an avalanche will be destroyed. Not only does the swift rush of air endanger any person or animal in the path, the snow dust will be inhaled, the particles will promptly melt in the lungs, and the victim will be asphyxiated precisely as though he had been drowned. This type of avalanche is particularly dangerous because it may occur on any steep slope with little or no warning at almost any time of year. Although it may leave its mark on a snowy mountainside, it has almost no effect as an agent of erosion, however much destruction of property and life it may entail.

In contrast, "wet" avalanches, which are essentially snow slides, contribute significantly to land sculpture. Although the sliding snow itself may move at more than a mile a minute, such avalanches do not ordinarily involve clouds of snow moving at hurricane speed. They are sometimes called "ground" or "debris" avalanches since they generally include all the earth and debris

216

beneath the snow, down to the solid rock. This type of avalanche occurs most commonly in the spring when water from melting snow or rain loosens a mass of close-packed snow and facilitates the downward pull of gravity. Generally the wet snow breaks into clumps which roll down, snowballing to huge size as they go, but sometimes great slabs of ice-hardened snow may hurtle over the slick surface of water-soaked banks. Such avalanches often occur year after year in the same ravines or gulches, developing a permanent slide and scouring it afresh each spring. Although generally far more massive than the "dry" avalanches, they are much more predictable both as to time and place and rarely catch the local inhabitants or experienced mountaineers napping. The path of many avalanches of this kind displays a *starting niche,* a *track,* and a *cone of deposition* on the mountainside. Thus they sculpture the land even as glaciers do, albeit on a smaller scale.

Avalanches ordinarily consist largely of snow and ice but they sometimes involve the movement of so much soil and rock debris that they closely resemble mudflows. This was true of the fearfully destructive avalanche that devastated a large area in the valley of Rio Santa in Perù on January 10, 1962. This agricultural valley, known as the Callejon de Huailas (Corridor of Greenery), is a thickly settled depression at an altitude of ten to twelve thousand feet in the shadow of the loftiest mountain in Peru, 21,834-foot Nevado Huascaran, with its perennial snowfields and garland of glaciers.

The avalanche started at the lower end of one of those glaciers. Weakened by warm weather, a mass of snow and ice estimated to have been about eight hundred yards long and two hundred yards wide with a volume of two and a half million cubic yards suddenly hurtled down the steep slope. Within two minutes it plunged into the populated valley of the Rio de Shacsha. By this time it had picked up quantities of soil, rock debris, and vegetation and had become more fluid as the ice was crushed and churned in the turbulent mass. Confined within the Shacsha canyon, it turned downvalley and fanned out to a width of about a mile on the broader floor of the Santa valley. There the churning mass of rock-strewn mud, with its fragments of ice, swept through towns and villages, demolishing everything in its path. The avalanche is reported to have descended ten thousand feet and traveled nine miles in less than ten minutes. The mile-a-minute speed permitted few to escape. The margins of the flow were so sharply defined that, in at least one village, many houses were buried under fifty feet of debris whereas neighboring buildings were left intact.

LANDSLIDES

Usually, the slope down which a landslide (in the restricted sense) moves is steep and the movement rapid. Usually, too, the moving mass has enough water to lubricate the slip surfaces, but if more than that is present, it is likely to acquire the character of a debris avalanche or mudflow. Depending upon the nature of the material before it begins to move downward, a landslide may be designated as a *debris slide* or a *rockslide.*

Debris slides are the most numerous and the most widespread. They result from downward movement of the unconsolidated materials that mantle the solid rock in many places—the products of weathering, the deposits of glaciers and their meltwater, the silt and clay deposited by streams, the artificial fill of "made land," and so forth. They often have a hummocky topography which may resemble that of a glacial moraine, but they can generally be identified by their relation to the scars that mark the spot from which they came. Such slides occur most frequently in undercut banks of meandering rivers, along shores undergoing vigorous erosion by waves, and on steep slopes in hilly or mountainous regions. They are usually small, but in the aggregate they are an important factor in wearing uplands down toward sea level.

There is soil and weathered rock debris in every rockslide but this type of landslide consists essentially of newly detached segments of the bedrock which slide downward on bedding planes, joint planes, or any other surface of separation. The mass of material is frequently so great that its momentum may carry it far beyond the foot of the slope on which it started. Landslides of this type are conspicuous features in many mountainous regions and along rocky headlands on many seashores. In populated areas they frequently result in catastrophic loss of life and destruction of property.

Such a rockslide occurred in the valley of the Gros Ventre River in Wyoming in 1925. The mountain ridge rising steeply above the valley floor at that point is composed of layers of sandstone, limestone and shale, all of which are inclined toward the valley at angles of eighteen to twenty-one degrees. A mass of some fifty million cubic yards of rock, mainly sandstone, suddenly slid downward on a layer of water-soaked clay shale, descending more than two thousand feet to crash on the half-mile-wide valley bottom. Its front plunged across the valley and rose 350 feet up the opposite side. Settling back, it became a dam 225 to 250 feet high; upstream a lake almost five miles long promptly formed. Much water seeped through the dam but there was danger that spring floods would cause the lake to overflow the top of the dam and breach it. This did not happen the first year, but in May 1927, water flowed over the dam and in a few hours the overflow cut a channel fifty feet deep, through which enough water rushed to cause a disastrous flood farther down the valley where several lives were lost despite warnings dispatched when the dam began to fail. After a few hours of rapid discharge of lake water, cutting was retarded by the greater width of the dam at the lower level and the outflow was reduced to more normal proportions. The relation between the lake, its outlet, and the rockslide dam has become fairly stable since 1928 and another disaster is unlikely in that valley.

In any such rockslide the moving segments of bedrock break up as they move downward. Hence the slide is conspicuous not only for the scar it leaves behind, high on the mountainside, but also for the commonly tongue- or lobe-shaped mass of crushed rocks spread on lower slopes or across valley floors. Eventually the scar will be smoothed down by weathering, while vegetation may obscure the heaps of rock fragments, but a trained observer can discern the forms that tell the story of even a geologically ancient slide.

Thus the rockslide that caused the disaster at Frank, Alberta, is easily recognizable today by the heaps of rubble below the scar, high on the side of Turtle Mountain (Plate 86). The mountain is composed of thick-bedded limestone overlying sandstone and shale in the midst of which are seams of coal. Prior to the landslide, the removal of coal, shale and sandstone in mining operations weakened the support of the overlying limestone. The

collapse of the mountainside came without warning. Great slabs of lime-stone slid downward and crashed in fragments; thirty-five to forty million cubic yards of debris spread across the valley. The momentum of the moving mass carried its forward edge four hundred feet up the opposite valley wall.

Both in the Gros Ventre Valley and at Frank, as well as at Elm, Switzer-land, the rock structure and the topography were conducive to the mass movements. The bedding planes of the sedimentary rocks in the Gros Ventre Valley, the joint planes in the limestone at Frank, and the planes of slaty cleavage and schistosity at Elm, all were potential slip surfaces inclined sharply down toward the base of a steep slope. At two of these localities, mining and quarrying operations removed part of the supporting rock. At all three, infiltrating water lubricated the slip surfaces and the expansion of ice in cracks contributed to the insecurity of the mountainside. Finally the force of gravity displayed its power in spectacular fashion.

Only a small minority of the earth's rockslides in recent geologic time

The massive "quick clay" landslide at Surte, Sweden in 1950, left a deep scar (shown at left), crossed a highway (center), and bulged far into the Gota River. The undisturbed region near the bottom of the photograph suggests how the entire area looked before the slide occurred. (Royal Swedish Geotechnical Institute)

219

have in fact been triggered by human activities, although the growing number of giant, earth-moving machines makes it possible that the percentage will increase. Usually natural processes are well able to set the stage without human aid. The normal downward erosion by youthful streams often leaves slopes too steep for the bedrock to maintain stability. Innumerable small rockslides and an occasional large one have contributed to the widening of many a valley above its stream channel. Erosion by valley glaciers characteristically steepens valley walls and many rockslides have occurred after the supporting ice has melted away during the waning stages of a glacial epoch. Undercutting of shoreline cliffs by wave erosion is obviously conducive to mass movements that may include rockslides. In all cases the structure and composition of the bedrock influence the location, dimensions and nature of the landslide.

An excellent illustration of the role of wave erosion in preparing the way for a rockslide can be seen along the California coast at Point Fermin, near Los Angeles. Here several suburban homes are located on a wave-cut terrace, with a highway running along the edge of a cliff that drops precipitously to the strand below. During storms the waves undercut the cliff and in the late 1950's a long segment of it slumped downward. Several hundred yards of highway dropped a dozen feet or more and houses were left standing on the edge of an abyss. The bedrock here is sedimentary; its fairly well consolidated layers incline gently toward the sea. The movement was outward as well as downward, primarily sliding along bedding planes. There was minor slumping into the gap at the rear of the main mass and a few small rockfalls from the cliff. But the momentum was insufficient to shatter the block or produce results even remotely resembling those at Elm or Frank.

Landslides have often been triggered by earthquakes. One such slide had catastrophic consequences in the valley of Madison River, some fifteen to twenty miles west of Yellowstone National Park, during the night of August 17, 1959. At that point the precipitous walls of Madison Canyon rise to heights of more than two thousand feet above the narrow valley floor. This portion of the Madison Range is composed of steeply inclined metamorphic rocks of Precambrian Age. For a couple of miles along the course of Madison River, a buttress of dolomite precariously supported a mountainside of deeply weathered schist and gneiss. Conditions were especially favorable for landslides. A major earthquake just before midnight violently shook the entire region. The vibrations shattered the dolomite buttress and eighty million tons of rock fragments came down from the mountainside in a roaring mass. At least twenty persons who were camping in the valley were killed. So great was the momentum of the debris that some of it crossed the canyon floor and climbed four hundred feet up the opposite wall. The landslide effectively dammed the river and a lake began to form upstream. Because of the grave danger that the waters of the "Madison Earthquake Lake" would break through the landslide debris and engulf the towns down the valley, channels were bulldozed through the dam to prevent the lake from becoming too deep.

The main line of the British Railways between Folkestone and Dover runs along the shore of the English Channel at the foot of the famous White Cliffs. At certain places those cliffs rise directly from the water's edge and the railroad, as at Shakespeare Cliff, goes through tunnels that parallel the

Above: The great landslide of August 17, 1959, in the canyon of Madison River, Montana. (Bureau of Reclamation)

Above, right: A typical avalanche of wet snow and earth in Switzerland. (Swiss National Tourist Office)

face of the cliff or cut through jutting headlands. Near Warren Halt the tracks are laid on a considerable strip of low hummocky land between cliff and shore. The hummocks are weathered blocks of rockslide debris from the chalk cliff. The jumbled mass is far from stable; repeated movements have displaced the railway, the sea wall built to protect the land from wave erosion, and the groins designed to deflect the longshore currents. Some of these movements were caused by the descent of wedges of chalk, newly loosened from the high cliff; others were due to subsidiary landslips—as the British call them—within the mass of slide debris.

Displacement of the railroad tracks is a serious matter and this large coastal landslip has been investigated in great detail by railway engineers and geologists. The massive beds of Cretaceous chalk in southeastern England are underlain by layers of marl, clay, and "greensand." At this point the uppermost layer of these older sedimentary materials is a few feet above sea level and all the beds are inclined gently toward the shore. When the clay and marl are saturated with ground water seeping downward from the uplands above the cliff, they provide a well-lubricated slip surface across which the overlying chalk moves easily. The railway engineers have tried to counter this and stabilize the slide by inserting drainage pipes in the clay layers to reduce the moisture along the slip surfaces beneath it. This has proved fairly successful.

ROCK GLACIERS

Great heaps of rock fragments catch the eye at certain places in many lofty mountains, such as in the Engadine region of the Swiss Alps and the San Juan Mountains of southwestern Colorado. The angular and little-weathered fragments, generally of volcanic or granitic rocks, closely resemble those of ordinary talus. The pieces, which range in size from sand grains to fifteen-foot blocks, are mixed haphazardly just as in the talus at the foot of a steep slope. But the shape of these heaps is quite unlike that of those evenly sloping piles of rock fragments. Moreover, they generally occur on the floors of cirques sculptured by glaciers during the Great Ice Age or on the higher slopes of peaks and ridges. Seen from some high vantage point, they appear like small glaciers, stretching outward and downward from the foot of a cliff. Or if they have a billowy surface, as many of them do, they may resemble a broad tongue of viscous lava flowing across the floor of a volcanic crater. The surfaces of such accumulations slope downward at angles of nine to eighteen degrees and generally display a pattern of rounded ridges and furrows parallel to the sides of the well-defined mass, becoming concentric at its forward end (Plates 84–85). Various names have been applied to features of this kind, such as "rock stream," "talus glacier," "coulées de bloc," and "glacier de pierres"; *rock glacier* seems to be the most graphic and suitable term.

Several theories have been offered to explain their origin, including the possibility that their well-developed longitudinal and concentric wrinkles are produced by rapid fragmental flow resulting from the shattering of rock-falls on impact. (Recall that the observers of the Elm rockslide said it "poured like a liquid"; this was doubtless true, but it was the flow of solid fragments, like the sand grains pouring through an hour-glass, not liquid flow.) It is more probable that most rock glaciers originally grew out of the disintegration by physical weathering of head walls and sides of glacial cirques left bare when the ice disappeared, with a few small rockslides thrown in for good measure. Then under appropriate conditions of topography and climate, a continuous slow flow produces the interesting surface features. This idea of slow flowage is borne out by observations on the rock glaciers in the Val Sassa and Val dell'Acqua in the Engadine Alps; movements at a rate of four or five feet per year have been measured there in the middle of the tongue-shaped masses, and somewhat lesser rates near their margins.

In the aggregate, the transfer of rock materials from higher to lower places by these innumerable mass movements is of great importance in the erosion of the land. The work done by gravity operating directly takes its place alongside the work of gravity operating indirectly through running water, blowing winds, and sliding glaciers. No estimates are available to rate the relative accomplishment of these several geologic agents, nor would any such estimates be worth while. In the last analysis, the energy of all of them is gravitational.

12

In the Realm of Vulcan

Of all natural phenomena the violent eruption of a volcano is the most spectacular and awesome. The gigantic clouds of sulfurous gas, steam and volcanic ash boiling upward from its summit are rivaled only by the mushroom cloud of a nuclear-bomb explosion. The incandescent blobs of molten lava hurled skyward, the pool of seething fire in the crater, the streams of lava pouring relentlessly down the mountainside—all come close to the usual conception of hell. It is no wonder that many peoples living near volcanoes have attributed volcanic activity to easily angered gods.

The ancient Romans wove legends and myths to explain in such terms the volcanoes of the Mediterranean region. They located the forge of Vulcan, the blacksmith god, beneath an active volcano on the small island of Vulcano in the Tyrrhenian Sea north of Sicily. At his anvil, Vulcan had fashioned Diana's arrows and Jupiter's thunderbolts, as well as the breastplate of Hercules and the shield of Achilles. The fire and smoke that occasionally belched from the mountaintop came from the chimney of his forge and the frequent subterranean rumbles and local earthquakes from hammer blows on his anvil.

Aside from Vulcano, the only active volcanoes the ancients knew were Stromboli, Ischia (which last erupted in 1302) and Etna. Vesuvius was not among these because for many centuries prior to A.D. 79 it was inactive; indeed, the ancient Greeks and Romans failed to recognize it as a volcano. Evidence available today indicates that Vulcano was more active than the others. Hence the Gran Cone on Vulcan's island is the prototype of volcanoes, and we salute the mythical god of the hearth and forge whenever we use the noun "volcano" or any of its derivatives.

VULCANO

Vulcano has not erupted since 1889 when from August to May a series of violent explosions produced a dense cloud of ash and cinders, and spread fragments of pumice, so-called "bread-crust bombs," and incandescent rocks on the flanks of the cone and over its immediate surroundings. Lava did not flow from the crater but the ruddy glow from the underside of the cloud hanging over the mountain indicated its presence. During earlier eruptions, such as those in 1775 and 1786, extensive flows of obsidian contributed to the construction of the great cone. Like most large volcanic mountains, it is a mixture of congealed lava and fragmental debris. At present the quiescent volcano rises to a height of 1266 feet. Its circular crater is about a quarter mile in diameter and five hundred feet in depth. Many fumeroles (vents that emit hot gas and steam) are occasionally active on the crater walls.

Halfway down the north slope of the Gran Cone is a small "parasitic crater" known as Forgia Vecchia. It grew there early in the eighteenth century when gas and lava broke through the side after the main vent was clogged. The far northern tip of the island has a separate, smaller cone known as Vulcanello, which apparently began to form during the first millennium B.C. Until the middle of the sixteenth century A.D. this cone made a tiny island separated from the larger island by a narrow strait; then the accumulation of ash joined the two together. Recent studies of the voyages and landfalls of Odysseus have led to the interesting suggestion that Vulcanello, a separate island in the time of Homer, was Charybdis, while Vulcano was Scylla.

STROMBOLI

Stromboli, located north of Sicily, is almost unique among the world's volcanoes in having been moderately active for more than two thousand years (Plate 101). When described by Pliny in the first century A.D., its general appearance and behavior were practically what they are today. Evidently the continuous escape of gas and steam from vents in its crater, the frequent minor explosions of incandescent ash, lapilli (fragments intermediate in size between ash and bombs), and bombs, and the occasional outpouring of small lava flows act as safety valves and prevent the storing of sufficient energy to cause a violent eruption. The lurid reflection of its ruddy glow from overhanging clouds may occasionally disappear, but rarely does a week pass without some display of its activity. Consequently, since ancient times Stromboli has been known as the "Lighthouse of the Mediterranean".

The mountain is an almost perfect cone rising from the sea to a height of three thousand feet; since the sea roundabout is approximately seven thousand feet deep, the cone is really ten thousand feet high, making it one of the world's largest volcanoes. Standing athwart the shipping lanes that lead from Marseilles, Genoa and Naples through the Straits of Messina, it is also one of the most frequently observed of all active volcanoes.

The tip of Stromboli's cone is truncated. The active crater is an oval basin whose nearly vertical walls are breached on the west side where a long slope, the Sciara del Fuoco ("ski of fire"), extends down to the sea. Only at intervals of a dozen years or so is there an eruption of sufficient violence to hurl ashes, cinders and bombs on the village of San Vincenzo on the eastern shore of the island; for this moderation the inhabitants may well be thankful for the breach in the crater.

VESUVIUS: FERTILIZER AND DESTROYER

Before A.D. 79, Vesuvius was a cone-shaped mountain with a broad, gently rounded summit four thousand feet above the head of the Bay of Naples.

224

Lush vineyards stretched halfway up its slopes. Strabo, writing about 30 B.C., reported that the cindery mass of its barren summit looked as if it had been "eaten by fire." Recounting the revolt of the Roman gladiators under the slave Spartacus about 72 B.C., Plutarch described what must have been a deep crater at its crest, with only a narrow opening on one side where Spartacus and his men took refuge for a time. Even so, the Roman inhabitants of the many villages and towns around the mountain seem to have been completely unaware of its volcanic nature until it erupted violently on August 24, A.D. 79.

The prosperous city of Pompeii, at the southwest foot of the mountain, was buried beneath fifteen to twenty-five feet of volcanic ash, much of it pumice ranging from particles the size of a pea to fragments two or three inches in diameter. The ash was the primary lava ejected from the newly opened vents and expanded by its own gas into a veritable froth which drifted downwind. The upper layers of ash were somewhat compacted by the torrential rain that typically accompanies an eruption. Excavation in recent years (Plate 94) has revealed a unique record of the life and customs of Roman times; casts of human bodies, utensils, furnishings and decorations were in a remarkable state of preservation.

The destruction of Herculaneum, a similar city west of Vesuvius, came about in another way. Instead of wind-blown ash it was overwhelmed by a mudflow that in places exceeded sixty-five feet in depth. This flow was a mass of ash, cinders, pumice and lava fragments washed from the upper slopes of the cone by the rain. Probably the mudflow descended like an avalanche and was of the type known to geologists as *lahar,* a slurry of hot volcanic debris that sweeps up everything loose in its path. The "mud" hardened as it dried and the consolidated material makes excavation much more difficult in Herculaneum than in Pompeii.

The eruption of A.D. 79 completely changed the appearance of Vesuvius, largely destroying the earlier dome-shaped summit and building a new cone of ashes and cinders. Repeated explosive eruptions occurred at intervals of fifty to a hundred years until 1036, by which time the mountain had assumed approximately its present form. All that remained of the ancient dome was the prominent ridge known as Mount Somma, forming a semicircle around the north and east side of Vesuvius. From 1139 to 1631 Vesuvius was in almost complete repose, and once more the vineyards spread up its flanks and prosperous villages dotted its slopes.

Then came the sudden catastrophic eruptions of December, 1631. Violent explosions in the crater threw up great clouds of ash and broken rock. Vast floods of lava issued from fissures high on the mountain slopes. Heavy rains started extensive lahars avalanching downward. Six towns were destroyed by lava and nine by lahars, and the ash falling on Naples was nearly a foot thick. At least four thousand persons were killed. The top five hundred feet of the great volcanic cone disappeared and the crater doubled in size.

Since that eruption, Vesuvius has been almost continuously active and its behavior has been so well documented that it is possible to discern cycles of activity, each covering fifteen to forty years. Each cycle begins with several years during which only a few jets of gas and steam rise from vents within the crater. After a time such emissions gradually become explosive and build one or more cinder cones on the crater floor. Lava may pour out

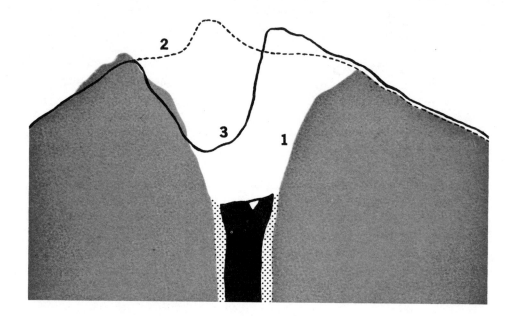

The summit of Vesuvius in three different periods: (1) after the violent eruptions in 1906; (2) in 1943 and in early 1944; (3) after the eruptions of March 22 to 29, 1944.

around the cinder cones and gradually fill the crater. This type of activity may continue for twenty years or so, but eventually the upper part of the conduit leading to the central vent becomes clogged with solidified lava and broken rock. Then come the culminating eruptions: the gas-charged liquid lava stands high in the throat of the volcano; pressure builds up until something must give; violent explosions cause local earthquakes and hurl great quantities of ash and cinders upward and outward; the sides of the cone rupture and floods of lava pour down. Paroxysmal eruptions continue for two or three weeks until the gas pressure has been relieved and a new cycle begins.

The last completed cycle covered thirty-eight years and culminated in the eruptions of March, 1944. When I flew over the crater in 1952, it was a funnel-shaped basin more than a third of a mile in diameter and several hundred feet deep. Not a whiff of steam or gas disturbed the gray and brown desolation of its interior; Vesuvius was in complete repose. The contrast with what I had seen in 1928, in the middle phase of the cycle, was most impressive. Then I had scrambled down from the crater rim to its flat floor only a couple of hundred feet below. There I walked gingerly over recently hardened lava crust, glimpsing through cracks the orange-red, still fluid lava beneath, and dodging the bombs occasionally lobbed from the summit of the cinder cone built up on the crater floor. The top of that small cone was almost exactly as high as the crest of the surrounding crater wall and from it a plume of fine ash, roaring steam and noxious vapor swirled continuously upward. Evidently the "gas blow-off" toward the end of the 1944 eruption had completely cleared the throat of the volcano before my second visit.

The slumber of Vesuvius from 1944 to the present (1963), poses an interesting question. Is the volcano in an extraordinarily long phase of repose (at least nineteen years to date, instead of the usual three to seven years), and if so, will it erupt violently within the next few years? Or has the sequence of cycles that began in 1631 reached its end and is the volcano now entering another long dormant period like that between 1139 and 1631? There is no

226

scientific basis for this prediction; only time itself can give us the answer.

The Vesuvian cycle is fairly typical of hundreds of active or recently active volcanoes in various parts of the earth. Moreover, thousands of extinct volcanoes now more or less eroded apparently were built by similar spasmodic eruptions. It is likely that any prominent, steep-sided volcanic cone grew in the same way as Vesuvius. The chemical composition of the lavas and other ejected materials ordinarily falls within the range of rhyolite and latite, two of the igneous rocks described in Chapter Three. Evidently the magmas feeding the volcanic vents were rich in silica, alumina, calcium, sodium and potassium, but comparatively poor in iron and magnesium. Consequently they were more viscous than the magmas responsible for diabase and basalt, which have a higher percentage of the latter elements. The greater the viscosity of a magma, the more difficult is the escape of the gases in it. This makes it more likely that the gas pressure will build up to a violent eruption. The difference in viscosity of magmas is the major cause of the contrast between the Vesuvian type of volcanic activity and, as we shall see, the Hawaiian type.

OTHER MEDITERRANEAN VOLCANOES

Mount Etna, one of the world's largest active volcanoes, rises to a height of 10,625 feet from a broad platform of sedimentary rocks about a thousand feet above sea level in the eastern part of Sicily. Eruptions from more than two hundred secondary or "parasitic" cones within a radius of twenty miles of its summit have made its sprawling mass more like a dome than a cone. The huge pile includes innumerable lava flows alternating with layers of material erupted in explosions; presumably there are in addition many dikes and pipelike bodies of intrusive igneous rocks beneath the surface. Weathering has formed fertile soil, and vineyards and cultivated fields spread around the lower slopes, with pastures at higher altitudes. Towns and villages are scattered from sea level halfway up the mountainsides. Several have been destroyed more than once during the Christian era, only to be rebuilt when the lava flow congealed or the lahar cooled.

Records of volcanic activity by Etna begin about 800 B.C. with a reference by Homer. There is also a description of an eruption in Virgil's *Aeneid*. Violent outbursts occurred during the Middle Ages but the greatest eruption came in 1669. It destroyed the summit cone and left a gigantic crater that now has several vents almost continuously emitting columns of gas and steam. The vast quantities of lava that flowed out from a fissure on the mountainside in 1669 destroyed the seashore city of Catania, ten miles away.

Another spectacular volcanic manifestation is in the Solfatara, near Pozzuoli five miles west of Naples. This is a large, flat-floored crater not far above sea level, one of nineteen craters comprising the "Phlegraean Fields" of the ancient Romans. The area can be reached by a good highway from Naples and includes Lake Avernus and a modern race track, each in an ancient volcanic crater, readily recognizable by its low rim. The Solfatara last erupted in 1198, releasing a stream of lava that reached the sea near Pozzuoli. Since then gas has risen constantly from several fumaroles near the circumference of the crater floor. Escaping from certain fumaroles with a noise like a locomotive blowing off steam, the invisible gas is so hot that it instantly ignites a paper roll thrust into it at arm's length. Other vents contain black mud constantly stirred by the gas bubbling through it and frequently spurting upward like a fountain. The gas is sulfurous and the name "Solfatara" is derived from the Italian word for "sulfur mine." Thus it is now customary to refer to any volcano from which only gas is emitted as being in the "solfataric stage."

Eastward from the southern tip of Greece, in the entrance to the Aegean Sea, is another famous volcano, Santorin. Actually the name applies to a small group of islands that enclose a nearly circular bay some six miles in diameter. The islands are the remnants of a large volcanic cone that collapsed or exploded in about 2000 B.C. and the bay is the largest submerged caldera (resembling a crater but several times as large) thus produced. Within the Santorin caldera lie three small islands, each the upper part of a volcanic cone built upward from the caldera floor long after the destruction of the ancient large volcano.

CIRCUMPACIFIC RING OF FIRE

Many active volcanoes, or volcanoes active in geologically recent time and now either dormant or extinct, are located within a few score miles of the shore of the Pacific Ocean. It is therefore sometimes said that the basin of that largest of all seas is surrounded by a "ring of fire."

Beginning on the curving arc formed by the Aleutian Islands and the Alaska Peninsula, more than thirty volcanoes have been active, at least to the extent of blowing a plume of gas and steam from their summit craters, in recent decades. Far to the west is Kiska, a symmetrical cone of intermingled lava and pyroclastic materials rising steeply from the water's edge. Midway in the island chain is Bogoslof which for ages has fought a battle with the sea; domes of lava have been squeezed upward and cones of cinders have risen above sea level only to be rapidly eroded by the vigorous waves or destroyed by explosions of volcanic gases. Reports of the appearance of new islands and the disappearance of old ones that sailors and aviators have occasionally brought back from this infrequently visited locality are really not exaggerated.

Farther to the northeast is Shishaldin on Unimak Island and near the base of the peninsula are Trident and Katmai, as well as Nova Rupta in the Valley of Ten Thousand Smokes. Katmai exploded with great violence on June 6, 1912, and spread vast quantities of ash and rock fragments over a large area. The Katmai ash accumulated to depths of two to three feet on housetops and in the streets of Kodiak, eighty miles to the southeast, and was identifiable in dust that fell on Juneau, more than seven hundred miles away. Trident was in moderate explosive activity in 1953 and again in 1961, but Katmai has been quiet since 1912.

The Valley of Ten Thousand Smokes achieved its name as a result of the eruption of a volcano born at about the same time as the 1912 eruption of Katmai. Nova Rupta is a low, gently sloping cone constructed largely of pellets and cinders of pumice on the valley floor. For seventeen miles below the volcanic vent, the valley is partly filled with a lahar in some places four

miles wide; from it the "ten thousand smokes" arise. When I persuaded my nervous horse to carry me up the surface of the lahar in 1923, we had to thread our way among many thousands of noisy jets of steam, but in the 1950's I was told by Alaskan Rangers that most of them had completely disappeared.

Proceeding clockwise around the Pacific border, there is the 14,000-foot-high summit of Mount Wrangell in the St. Elias Range north of Valdez, Alaska, displaying a brave column of gas and steam. Next is the Cascade Range extending from British Columbia into northern California. Its higher summits are volcanic cones reared by a Vesuvian-type of activity above the sedimentary and metamorphic rocks forming the "backbone" of the range. Active in late Tertiary time, most of these volcanoes became extinct before the close of the Pleistocene ice age. The extent to which erosion has modified the original conical or dome shape gives an idea of how long it has been since the volcanic fires were quenched. Each peak bears the marks of glacial erosion—cirques and ice-shaped valleys—and the loftier ones still have perennial snow fields and girdles of alpine glaciers. They include Mt. Baker, Mt. Rainier and Mt. Adams in Washington, Mt. Hood in Oregon, and Mt. Shasta in northern California. Shasta still shows vestiges of its former vol-

Vesuvius in violent eruption in 1944. The crest of Mount Somma, left foreground, is mantled with fresh snow.
(U.S. Air Force)

229

canic activity in the faint wisps of sulfurous vapors that occasionally issue from fumaroles near its summit.

A prominent feature of this range of recently extinct volcanoes is Crater Lake, Oregon, on the crest of the Cascades. The lake, about six miles in diameter, lies within a rim of precipitous cliffs from five hundred to two thousand feet high. Its marvellously blue water is two thousand feet deep at certain points. The basin is one of the most perfect calderas in the world. Throughout Pleistocene time, the spot was occupied by a huge volcano resembling Hood or Shasta, to which the name Mount Mazama has been given. During the Wisconsin glacial stage, glaciers descending its flanks left their marks on the outer slopes of the ridge that now encircles the lake. Some six to ten thousand years ago, Mazama exploded with violence rivaling that of Krakatoa in 1883. The blast presumably demolished the top of the cone. Then the greater part of the mountain collapsed into the magma chamber which had been emptied by gas expansion and by the settling of the magma. These events changed the crater into a caldera. Craters are the vents through which lava and pyroclastic material erupt to form cones; the largest of them rarely exceed a mile in diameter. Calderas result from a subsidence into an emptied or partially emptied magma chamber; they may be many miles in diameter. Soon after a caldera is formed, one or more new volcanic cones usually rise within it by eruption of the residual magma. Wizard Island in Crater Lake is just such a youthful cone with a typical·crater on its summit.

Some seventy-five miles southeast of Mt. Shasta stands Lassen Peak, reputed to be the only active volcano in the continental United States. It is farther from the Pacific shore than the extinct volcanoes of the Cascade Range but it must be considered within the ocean's "ring of fire." Its moderately steep cone, deeply scarred by erosion, rises above its surroundings only a little less than Shasta, Hood and Baker. Until 1914 it was classed along with them as an extinct volcano, but beginning in May of that year it produced a series of moderate eruptions, some involving the outpouring of lava as well as ash and cinders. Since 1915, except for a small outburst in 1925, only small plumes of gas and steam have issued from it; perhaps before long, Lassen will again be listed as an extinct volcano.

There is a wide gap in the "ring of fire" between the Cascade Range and the quiescent or currently active volcanoes in and near the Sierra del Sur in Mexico south of the Tropic of Cancer. The most northerly of these is Ceboruco in the State of Nayarit, last reported in eruption in 1875. Next in line are the Volcán de Colima and Fuego, which have been active in recent years. Toward the southeast, in the State of Michoacán, is Parícutin which has the distinction of having its birth observed by dozens of geologists.

Here, in the late afternoon of February 20, 1943, a farmer named Dionisio Pulido was startled by smoke or fine ash-gray dust rising from a fissure in his cornfield, accompanied by "a thunder," the odor of sulfur, and a loud hissing. When he returned the next morning from his home in Parícutin a mile and a half away, he saw a cone about thirty feet high, with "smoke" boiling upward from it and "stones" flying out. The cone grew to a height of 150 feet by nightfall. The next day, a slaglike mass of jagged blocks of black lava, about fifteen feet thick, spread out from the base of the cone and covered most of his farm. Later, Pulido found solace from his distress in the fees he received as guide for thousands of sightseers. The newborn volcano

View of Parícutin in Mexico, soon after it appeared in 1945. Tarascan Indian houses in the foreground have been demolished by the heavy ashfall. (Tad Nichols: Western Ways Features)

230

reached its climax of activity during the summer of 1943, when the cone was about nine hundred feet high. For a brief time in June, incandescent lava broke through the crater rim and cascaded down one side of the cone, but at all other times the lava issued from vents near the base. By the end of October, 1944, lava flows covered fifteen square miles and destroyed the town of Parangaricutiro, two miles distant, as well as the nearby village of Parícutin. The cone was 1250 feet high when the eruptions ceased as abruptly as they had begun. Since March 4, 1952, Parícutin has been "dead." It now takes its place among the dozens of geologically recent but apparently extinct volcanoes in that area.

Far surpassing in grandeur this and other recently born, short-lived volcanoes is Popocatepetl (Plate 103), visible to the south from Mexico City on a clear day. Its massive, snow-capped cone rises to a height of 17,887 feet and was built by eruptions continuing through several geologic epochs. A column of gas and steam frequently rises thousands of feet from a summit crater more than half a mile in diameter and five hundred feet deep. Farther to the east is Orizaba, more correctly known as Citlaltepetl, the highest (18,696 feet) and most beautiful of Mexican volcanoes. The almost perfect symmetry of its slender, steep cone and the gleaming purity of its lofty snow fields have only to be seen to be admired.

From Mexico southward through Central America and down the full length of South America to Cape Horn, the chain of volcanoes is practically unbroken. It includes in Guatemala, Santa Maria, which erupted early in the twentieth century after a long slumber, and Atitlán with its slightly eroded, symmetrical cone, as well as Coseguina in Nicaragua and Izalco in San Salvador (Plates 89 and 100). Farther along are the many volcanoes of Colombia, Ecuador and Peru, several rising high above the general summit level of the Andes. Among them are Chimborazo, reaching more than four thousand feet above its base to a summit elevation of twenty thousand feet and crowned with snow and ice despite its location almost on the equator, as well as Cotopaxi and Pichincha. And in Peru there is El Misti with its startlingly beautiful, symmetrical cone and its plume of white vapor rising from its occasionally active crater. Farther south in Chile, Riñinahue on the flanks of the larger Puyehue erupted explosively on May 23, 1960, and poured out a small lava flow. This occurred during a major earthquake that devastated central Chile and is one of the very few eruptions closely associated with known crustal movements of deep-seated origin, although such association is inferred for many volcanic outbursts in regions where mountain-building movements of the earth's crust have taken place.

The widest gap in the circumpacific "ring of fire" appears to be between the southern tip of South America and New Zealand, far to the west, but the ring is probably represented here by submarine volcanic activity responsible for many of the ridges and seamounts on the floor of the South Pacific hundreds of miles north of Antarctica. It reappears from beneath the sea along the curving axis extending from the southern tip of New Zealand's South Island to the northern tip of the North Island. Strung out along this curve, especially on the North Island, are many volcanic mountains: snow-capped Ruapehu, near the center of the North Island, from which lava exuded in 1945; not far from it, the still-growing cone of Ngauruhu with an active crater at the top; beyond it Tongariro and, some eighty miles eighty miles to the west of Ruapehu, the splendid cone of Mount Egmont rising high above snow line and overshadowing a smaller parasitic cone on its flank.

An extraordinarily violent series of explosive eruptions in 1886 produced well-nigh unique results in the Tarawera region, about sixty miles northeast of Ruapehu. Craters and elongated pits were blown out to form a nearly continuous trench about nine miles long, averaging an eighth of a mile in width and varying in depth from three hundred to fourteen hundred feet. The celebrated hot-spring deposits, White Terrace and Pink Terrace, were destroyed but a new geyser was born—the giant Waimangu which for several years hurled a column of muddy, steaming-hot water to a height of more than a thousand feet. Its activity ceased early in this century and the Tarawera line of craters is now quiet.

Conspicuous in the Bay of Plenty is White Island, the upper part of a volcanic cone, now in the solfataric stage, and in far northern New Zealand are such cinder cones as Te Ahuahu, four hundred feet high, surrounded by lava flows. Far to the north and northeast, the circumpacific ring reappears in New Caledonia, the New Hebrides, and the Solomon Islands, where the composite cone of Sava has occasionally been reported to be active. Just beyond, in New Britain, a new volcanic cone called Vulcan grew near Rabaul by explosive eruptions in 1937.

From New Britain, the chain of volcanoes curves sharply to the west to traverse New Guinea and then northward again in the Philippine Islands, where volcanoes are among the most conspicuous features of the landscape. The erosion of most of them indicates that they have long been extinct, but some, like Camiquin del Sur, Taal, and Babuyan Claro, have been in moderately explosive activity in recent centuries. So too has Mayon, a strikingly beautiful, nearly perfect cone towering upward more than eight thousand feet on the island of Luzon.

Continuing northward through Taiwan (Formosa), the chain curves eastward to follow the Ryukyu Islands and to extend the full length of the Japanese archipelago. Among Japan's score of active or recently active volcanoes are Sakarashima on the island of Kyushu, which erupted violently in 1914, and farther north on the same island, Aso-san with a caldera (on its mile-high summit) larger than that occupied by Crater Lake. On Honshu Island are Fujiyama and Asama. The latter is almost continuously in moderate activity, but Fuji rears its benign cone to a height of 12,388 feet in artistic symmetry, and its crowning filagree of snow has not been disturbed by ash or lava since 1707. The Japanese climb to the summit and look down into its five-hundred-foot-deep crater in quasi-religious pilgrimage; tourists from overseas are generally content to admire its beauty from afar.

Farther toward the northeast the circumpacific chain includes another score of active or recently active volcanoes among the Kurile Islands and on the Kamchatka Peninsula. Of those in the latter locality, Klyuchevskaya, Skopa and Bezymianny are the most notable. The outburst of the last named in 1956 was by far the greatest eruption of the twentieth century. Violent explosions and vast glowing ash flows (lahars) destroyed the entire summit of its mountainous cone in a process apparently like that believed to have made Crater Lake from Mount Mazama. It was a spectacular justification of the idea, firmly held by geologists, that all the features of the earth are explainable in terms of natural processes that continue today.

233

Branching toward the west and northwest from the ring of circumpacific volcanoes in the vicinity of New Guinea, the Indonesian volcanoes are strung out in a long arc that eventually curves northward and ends on Barren Island in the Bay of Bengal. This chain includes the many volcanoes on the islands in the Banda Sea, the Lesser Sunda Islands, Java and Sumatra. The majority of them are now extinct or have long been quiescent, but more than a dozen are active, some in the solfataric stage and some with intermittent, moderately violent explosions accompanied by outpouring of lava.

Most famous of Indonesian volcanoes is Krakatoa, an island in the strait of Sunda between Java and Sumatra. Its great eruption in 1883 gave much impetus to the science of volcanology. Prior to that year three composite cones of lava and fragmental material had grown upward from the sea floor until they coalesced to form one large island with three summits, 2700, 1460 and 400 feet high. Then in August 1883, a series of tremendous explosions heard nearly three thousand miles away produced black clouds of ash rising to heights of approximately fifty miles. The finer dust from the ash clouds traveled around the earth many times and caused sky glows observed in Europe and the United States throughout the fall and winter of that year. It has been calculated that five cubic miles of material was blown into the air, almost all of it new lava sprayed upward from the magma chamber by gas expansion. After the eruption, two-thirds of the island disappeared; two of the three cones had collapsed into the emptied chamber and only a fraction of the third remained on the rim of the submarine caldera thus formed. This is the island now known as Rakata.

All was quiet at Krakatoa from 1883 to 1927 when eruptions began again with submarine construction of a new cinder cone on the floor of the caldera. By 1952 its summit was 220 feet above sea level. The small island was promptly named Anak Krakatoa—the "child of Krakatoa." Eruptions continued to increase its size as recently as 1959.

MID-PACIFIC VOLCANOES

Probably the most massive range of volcanic mountains on earth stretches sixteen hundred miles across the central Pacific from Kure (Ocean) Island at the northwest to Hawaii at the southeast. Four-fifths of this range is represented on the map only by tiny dots of low islands and reefs, but these are the summits of submarine mountains that rise fifteen thousand feet or more from the ocean floor. The exposed parts of Kure and Midway Islands are composed entirely of coral limestone which, at the Midway air base, glares painfully white in the dazzling light of the subtropic sun. But at a shallow depth the limestone rests on the wave-eroded, truncated crest of broad-based volcanic cones. At French Frigate Shoal a small pinnacle of volcanic rock rises above the limestone reefs, and both Necker and Nihoa Islands are obviously the dwindled remnants of larger volcanic islands, long battered by eroding waves.

Farther to the southeast the largely submerged mountain range achieves grandeur in the eight major islands of the state of Hawaii. Each of these is

the more or less eroded summit of a volcano or cluster of volcanoes. Only on the largest and most southeasterly of these islands, Hawaii itself, are the volcanoes active at present. Evidently the long, slightly curving line of volcanic vents marks the location of a major fissure or zone of weakness in the earth's crust. Lava first welled upward at the northwest end and shifted progressively toward the southeast when the accumulation of volcanic debris sealed the northwesterly conduits. The basaltic lava is far more fluid than the viscous lava typical of Vesuvian-type volcanoes. Consequently the successive flows have built broad domes called "shield volcanoes," rather than steep cones.

Oahu, on which Honolulu is situated, is the deeply eroded summit of two of these shield volcanoes, Waianae and Koolau, which long ago coalesced to make a single island. Neither has been active in historic time and the extent of the erosion makes it safe to classify them as extinct. Diamond Head is a secondary cone, related to the last eruptions of Koolau. The many lava flows that accumulated one above another during its construction are ten to twenty feet thick and some are clearly visible in the wave-cut cliff facing Waikiki Beach.

Hawaii, the "Big Island," is the upper part of five coalesced shield volcanoes. Its twin summits, Mauna Kea and Mauna Loa, rise 13,823 and 13,675 feet respectively above sea level and approximately 30,000 feet above the

A grove of ohia trees killed by a fall-out of ash and cinders from the eruption of Mauna Iki in Hawaii in 1960. (Tad Nichols: Western Ways Features)

235

ocean floor. This huge pile contains about ten thousand cubic miles of erupted material as compared with the eighty cubic miles in the volcanic cone of Mount Shasta. Of the five volcanoes in this cluster, Mauna Kea and Kohala have been inactive for many thousands of years and are presumably extinct. Hualalai has been dormant since 1801, but Mauna Loa and Kilauea continue to provide magnificent eruptive spectacles every few years.

At the summit of Mauna Loa is an oval depression three miles long and a mile and a half wide, known as Mokuaweoweo. It is locally called a crater but is actually a caldera whose depth depends upon the ratio between the amount of new lava that occasionally floods its floor and the subsidence during or after eruptions. It is generally a few hundred feet deep. For the last fifty years or more, eruptions have alternated between lava fountains and flows in the caldera and on the flanks at irregular intervals of a few months to several years. The eruptions of January, 1949, from the caldera, and of June, 1950, from a fissure about seven miles long, were especially spectacular.

Kilauea Volcano is a separate lava dome built up against the southeast side of Mauna Loa to a height of 4090 feet above the sea. Near the southwestern edge of the floor of its caldera is a circular pit known as Halemaumau, the "House of Everlasting Fire," which, prior to the explosive eruption of 1924, usually contained a boiling lake of liquid lava. This may be the place at which the principal lava conduit reached the surface, but much of this huge shield volcano was constructed of lava poured out from craters and fissures in the rift zones that extend down its flanks eastward and southwestward from the summit caldera. The eruption of Kilauea in 1952 ended the longest period of inactivity in its historic record—nearly eighteen years. Then in 1955 about a hundred and twenty million cubic yards of lava poured out from numerous vents along the rift zone extending eastward and northeastward. Awesome fountains of lava rose six or seven hundred feet above several of the vents, and three of the lava flows streamed down into the sea. Nearly five years later, in the winter months of 1959-1960, activity was resumed at both ends of that rift zone. Kilauea-Iki, a pit immediately adjacent to Kilauea Caldera, contained a lava lake that at one time during this eruption was four hundred feet deep, and the village of Kapoho (Plate 99), nearly thirty miles away, was destroyed by flows that eventually reached the sea and extended the shore outward as much as half a mile.

89. Izalco Volcano, San Salvador, shown at night during a spasm of moderate activity in January, 1957.
(John W. Mulford)

90. Overleaf: Night view of the crater of Niragonga, Congo, one of Africa's most active volcanoes.
(Emil Schulthess)

IN THE WEST INDIES

No "ring of fire" surrounds the Atlantic Ocean basin; its geologic structure and history differ significantly from that of the Pacific. Nevertheless, the 450-mile-long "island arc" of the Lesser Antilles is a volcanic chain at the edge of the basin. This portion of the West Indies is a curved line of small islands extending from the Anegada Passage, east of the Virgin Islands, to Grenada near the coast of eastern Venezuela. The Leeward Islands at the north and the Windward Islands at the south are the two parts of an arc bowed out toward the Atlantic and separating it from the Caribbean Sea. Each island is the upper part of a more or less eroded volcanic cone, or cluster of cones, standing on a broad, flat-topped submarine ridge that rises three thousand feet or so above the sea floor but still is five thousand feet

93. Aerial view of the summit of Ubinas, Peru.
(Fred M. Bullard)

91. The 1912 eruption of Mt. Katmai, Alaska, produced
a huge crater in which Emerald Lake and a small glacier
(above, left) may now be seen from the air.
(Steve McCutcheon)

92. Within the last few thousand years, volcanic explosions
in Ethiopia have produced immense craters (left) now
filled with water.
(Laurence Lowry: Rapho Guillumette)

96. The latest lava flow in the MacKenzie Lava Field,
Oregon, is so recent that the scrubby vegetation on the older
flows has not yet spread across it. (Josef Muench)

94. In 1954, Via di Stabia (above, left) separated the
excavated part of Pompeii from the unexcavated region on
the right. Vesuvius is in the background. (Douglas P. Wilson)

95. Some cinder cones and lava flows in the Craters of
the Moon National Monument (left), Idaho, result from volcanic eruptions in the last thousand years. (Josef Muench)

97. Many of the older lava flows (above, left) in Iceland have been deeply entrenched by stream erosion, and in some places active fumeroles may be seen along the canyon walls.
(Alfred Ehrhardt Film)

98. The explosive eruption that formed this volcanic crater (left) in Iceland took place so recently that its walls have scarcely been marked by erosion.
(Alfred Ehrhardt Film)

99. The Kapoho eruption of Kilauea (right) in early 1960 ejected thousand-foot fountains of lava, including ash and pumice that covered roads and crumpled roofs.
(Willard Parsons)

100. Volcan Izalco, San Salvador, was unusually active in January, 1957, when lava flows broke through its flanks and great clouds of gas and ashes rose high above its summit.
(Frederick H. Pough)

101. In 1952, the Sciara del Fuoco, or Scar of Fire, extended all the way from the summit of volcanic Stromboli to the Mediterranean Sea. (Fred M. Bullard)

102. Snow and ice crown the summit of Mt. Kilimanjaro in striking contrast to the tropical flora and fauna of the surrounding lowland. (James R. Simon)

103. Seen from the vicinity of San Martin, the great Mexican volcano Popocatepetl (left) rivals Fujiyama in the symmetrical beauty of its snow-capped cone. (Mary S. Shaub)

below sea level. The volcanic cones therefore rise eight to ten thousand feet above their bases. Each has been active in fairly recent geologic time, but only two—Pelée on Martinique Island and La Soufrière on St. Vincent—have erupted in the last few centuries.

THE GREAT ERUPTION OF MOUNT PELÉE

The eruption of Mount Pelée in May, 1902, one of the greatest catastrophes in recorded human history, killed more than forty thousand persons and destroyed the entire city of St. Pierre. Although several moderate explosive eruptions had occurred in the ancient caldera at the summit during the first few days of that month, no one foresaw what was to happen on the morning of May 9th when the volcano literally exploded and shot out a tremendous glowing cloud of superheated steam and red-hot dust and ash. Part of the cloud was heavier than air, in spite of its high temperature, because of its large content of dust particles, and this caused it to sweep down the mountainside close to the ground at hurricane speed. St. Pierre lay directly in the path of the incandescent avalanche and all but two of its inhabitants were killed almost instantly. Intermittent eruptions of a similar type continued for several months and some of them were watched from a safe distance by the scientists who rushed to the scene from the United States and Europe. It was a type of eruption not previously observed by volcanologists but now well understood. The glowing clouds are called *nuées ardentes;* they are formed by the violent explosion of viscous lava and generally are not accompanied by the outpouring of lava flows. They involve a magma rich in silica and alumina, of the kind that would solidify as rhyolite or latite, not the more fluid magmas of the kind that would solidify as basalt. Between a *nuée ardente* and a lahar there is complete gradation, depending upon the ratio between the total volume of solid particles and of the gas or steam. The former is a cloud of gas in which the solid particles are suspended; the latter is a mass of ejected fragments kept mobile by associated gas, steam and heated air.

By October, 1902, the ejection of *nuées ardentes* from the crater of Mount Pelée had practically ceased. The lava in the volcano's throat had lost most of its gas content and a plug of solid or near-solid lava formed like a stopper in a bottle. Then another strange phenomenon appeared. A slender tower or "spine" began to rise above the mountaintop; the "stopper" was being pushed upward by the still expanding liquid lava in the conduit below. Pelée's "spine" reached to more than a thousand feet above the crater floor by mid-1903, but thereafter it crumbled away more rapidly than it rose.

Pelée then remained quiescent until new eruptions in September, 1929, which continued for nearly three years. They were similar to those of the 1902-1904 period, but somewhat less violent. Many *nuées ardentes* were emitted and quantities of ash fell from turbulent clouds that rose many thousands of feet. Again no lava flowed, and the eruption came to an end when the vent was sealed by a dome of solid lava that partially filled the crater alongside the stump of the 1903 spine. Pelée has been dormant since 1932; whether it will erupt again cannot be foretold. It "slept" for twenty-five years between 1904 and 1929.

104. Shiprock, New Mexico, is the eroded remnant of a lava plug that congealed long ago in the throat of a volcano that has been almost completely worn away by water and wind. (Josef Muench)

249

Many volcanologists believe that this Peléan type of eruption signals the approaching extinction of an old volcano. Certainly it is not the kind of activity that built Mount Pelée, a huge cone constructed by innumerable eruptions of the Vesuvian type. That belief may be shattered by renewed activity at any time. If so, there will be premonitions in the form of local earthquakes and minor ejections of ash-darkened plumes of vapor from the mountaintop to warn inhabitants of imminent danger. If not, another century or two of complete inactivity will be required to confirm it.

ERUPTION OF LA SOUFRIÈRE

Closely similar to Pelée in size and composition is La Soufrière, a volcano forming the north part of St. Vincent. This island is ninety miles south of Martinique, and at its 4048-foot summit are two craters. One of them erupted in 1812; the other is the so-called "old" crater, from which violent eruptions were practically contemporaneous with the disastrous explosions of Pelée in 1902. This may well have been a mere coincidence; there are good reasons for believing that no effective subterranean connection existed between the two. The eruptions from the two volcanoes were quite similar. Relatively more ash was exploded from La Soufrière and some of the ejected material was much coarser; red-hot bombs, six inches in diameter, fell in Georgetown five and a half miles from the crater. Unlike Pelée, La Soufrière has been dormant since 1902. There is, however, no evidence of a lava plug sealing the supply pipe on its crater floor, and it should therefore be considered capable, at least for another half century, of renewed destructiveness.

VOLCANOES IN MID-ATLANTIC

In the Atlantic, as in the Pacific, many volcanic mountains rise high above the deep sea floor. All the rocks exposed in the islands of Bermuda, 580 miles east of the coast of South Carolina, are limestone, much of it wind-laid coral sand. This is a mere veneer, three hundred feet or so in thickness, on the truncated summit of an early Tertiary volcano at least six thousand feet high. The Cape Verde Islands, some three hundred miles off the coast of Africa, are the more or less eroded summits of a cluster of even larger volcanoes. The Pico do Cana, on the island of Fogo (the Portuguese word for fire), rises almost ten thousand feet above sea level and erupted rather violently several times in the eighteenth and nineteenth centuries. "Smoke" and ashes from its summit crater can frequently be seen far out at sea.

The most astonishing feature of the Atlantic basin is the Mid-Atlantic Ridge. The contours of this great submarine structure were little known before the beginning of the International Geophysical Year in 1957. It is an almost completely submerged mountain range, two to three miles high, that extends in a sinuous line from Iceland at the far north to the Tristan Da Cunha Islands and beyond them at the far south. At widely separated distances along this "ridge," a few of the many volcanoes associated with it rear their summits above the sea.

Iceland is essentially a volcanic plateau constructed of innumerable sheets

A fissure in lava flows in Iceland is typical of the fractures that often form after an eruption. (Alfred Ehrhardt Film)

of basaltic lava that poured out from fissures several miles in length and spread far out on either side. The plateau rises several thousand feet above the sea floor and is the northern extension of the Mid-Atlantic Ridge. Fissure eruptions have continued in historic time; in 1783, a lava flow from the Laki fissure covered an area fifty miles long and up to fifteen miles wide, and another flow a little later in that year was forty miles long and in places seven miles wide. The average thickness of these sheets of lava was about a hundred feet. At several places on the island, the eruptions have been locally concentrated along a fissure, and shield-shaped or domelike volcanoes have grown there. The most famous of these is Mount Hekla, whose alternating sheets of lava and beds of ash rise three thousand feet above the plateau to a summit elevation of nearly five thousand feet. Its latest eruption was in 1947, following 102 years of quiescence, and involved the ejection of much ash and lava. Icelandic lavas are in general somewhat richer in silica and alumina than the typical basaltic lavas of Hawaii; this presumably accounts for the more explosive type of eruption and the ejection of greater amounts of fragmental debris.

Southward from Iceland, the crests of the Mid-Atlantic Ridge break the surface of the sea at only two places in the North Atlantic: the Azores, twelve hundred miles west of Portugal, and St. Paul's Rocks midway between the eastern bulge of South America and the great curve in the west coast of

Africa near the equator. The Azores are the summits of volcanoes, five of which have erupted in historic time. Fayal Island, for example, is the upper part of a Hawaiian-type shield volcano with gentle slopes rising to 3351 feet where a caldera, a mile and a half in diameter and nearly a thousand feet deep, completes the picture. A line of cinder cones marks a rift zone extending westward from the caldera to the sea at Capelinhos Point. Lava issued from near the middle of that rift in 1672 and poured far down into the sea. In 1957-1958, a new volcano rose from the sea, a mile offshore at the far end of the rift zone. At first it was composed only of cinders and was badly battered by the waves, but when lava flowed out from its flanks it became more resistant to erosion. At latest reports, Capelinhos stood five hundred feet above sea level and had tied itself to Fayal Island by its own eruptive materials.

St. Paul's Rocks are tiny, desolate islands, the dwindled remnants of the wave-beat summit of a volcano that became extinct in late Tertiary time. Far beyond them in the South Atlantic, widely spaced but still following the sinuosities of the Mid-Atlantic Ridge, are Ascension Island, St. Helena Island, the Tristan Da Cunha Islands, Gough Island and, near Antarctica, Bouvet Island. Each of these is the largely eroded summit of a huge, nearly submerged volcano or volcanic complex. The configuration of a great caldera can be seen on the summit of Ascension's Green Mountain at an elevation of 2820 feet, and Diana's Peak (2704 feet above the sea) is a remnant of the rim of an even larger caldera on St. Helena, but the slopes of both of these volcanoes are deeply trenched by ravines and youthful valleys radiating toward the sea. The volcanic cone on the largest of the Tristan Da Cunha Islands rises from its base, at least 6000 feet below sea level, to a summit of 7640 feet above the sea; it is therefore one of the major volcanoes of the

Eruptions of the new Capelhinos volcano in 1957 and 1958 greatly enlarged Fayal Island in the Azores. At times, great clouds of steam and ash (above, left) dwarfed the 115-foot-high lighthouse at the west end of the island. All was quiet when the photograph at right was made in September, 1960. (U.S. Air Force)

world. The crater at its summit has not been greatly modified by erosion, but cliffs a thousand feet high have resulted from the onslaught of the waves against its flanks since it became dormant. An eruption in October, 1961, made necessary the evacuation of all the islanders to Great Britain, much to their displeasure; however, by 1963 they were able to return to their austere homeland.

MIDCONTINENTAL VOLCANOES

Only a small minority of the world's active or recently active volcanoes are located in the interior of the continents more than a hundred and fifty miles from the sea. Outstanding among them are the several volcanoes of central and eastern Africa, the largest of which are closely associated with the great rift valleys of east Africa. The most notable is Kilimanjaro (Plate 102) in Tanganyika, Africa's highest mountain (19,590 feet), which in spite of its nearness to the equator is garlanded with snowfields and glaciers. It consists of twin volcanoes that are joined to a height of 14,000 feet but have separate cones seven miles apart above that level. The lower one has long been extinct, but in the crater of the other are small cinder cones constructed within recent centuries. To the north is Mount Kenya, rising to a height of 17,040 feet. Still farther north and near the southern end of Lake Rudolf, the volcano known as Teleki erupted as recently as the close of the nineteenth century and, far to the west in central Africa, Niragonga (Plate 90) is today one of Africa's most active volcanoes.

The midcontinental volcanoes of eastern Africa are definitely associated with major rifts in the earth's crust. So too are the extensive lava sheets of Africa's widespread plateaus and upland plains, the products of fissure eruptions in early Tertiary time. Such relations between crustal fractures and volcanic features are not apparent in the western interior of the United States. The vast plateau basalts in the Columbia and Snake River valleys issued from many widely scattered fissures, the precise location of which is unknown, and covered nearly two hundred thousand square miles in eastern Washington and Oregon and southern Idaho. The major eruptions ceased long enough ago to permit the carving of canyons two to four thousand feet deep; on their steep walls, lava flows ten to two hundred feet thick can be seen piled one above another. At several places, however, the uppermost flows have retained their pahoehoe or aa surface texture and are so little weathered as to give the appearance of having solidified only yesterday. At such places, too, there are generally many small cones of lava or of cinders standing scores of feet above their surroundings and scarcely modified by erosion. Evidently the waning volcanic activity continued locally well into historic time. One such locality, in southern Idaho, has been designated The Craters of the Moon National Monument (Plate 95).

Other evidences of geologically recent volcanism are widely spread in the states of Utah, Colorado, Arizona and New Mexico. The San Francisco Mountains, seventy miles south of the Grand Canyon in Arizona, are much-eroded volcanoes that still rise more than three thousand feet above the plateau on which the eruptive rocks were piled in Tertiary time. In the general vicinity, hundreds of cinder cones and lava flows show so little weathering and such slight etching by erosion that they must be less than two or three thousand years old. The last eruption at one of the cinder cones, Sunset Crater, has been dated by tree-ring chronology at about A.D. 1060.

The Spanish Peaks in south central Colorado are Tertiary volcanoes like the San Francisco Mountains, but even more deeply eroded. Erosion on the flanks of the western peak has uncovered a remarkable system of radiating dikes, some now standing as spectacular walls above their surroundings. These were fissures inside the volcano through which lava made its way from the central conduit toward the flanks of the cone. The last of the liquid lava solidified in them as the volcanic fires died. The relation of dikes to central conduit is even more completely revealed at Shiprock (Plate 104) in the Navajo Reservation in northwestern New Mexico. Here only the skeleton of the quondam volcano has successfully resisted destruction and removal by erosion. The "Rock" is the solidified lava that plugged the conduit when the volcano expired. The igneous rock of the volcanic plug (or neck) and dikes resists erosion far better than the ash and cinders of the cone and the sedimentary rocks on which it stood. Consequently these plugs and dikes remained as conspicuous features of the landscape after the cone was eroded away.

Volcanic activity was widespread and intense during Tertiary time in southern Colorado and New Mexico where many of the mountains are the result of erosion of huge piles of erupted fragmental materials interspersed with lava flows. The Jemez Mountains, northwest of Santa Fe, are of special interest because the Valle Grande which they encircle is possibly the largest caldera on earth. Within its 176 square miles are cinder cones such as El Cajete, the most recently active, which ejected quantities of "popcorn" pumice and poured out a lava flow that congealed as obsidian not more than two or three thousand years ago.

IMPACT CRATERS

In this same region is a feature of the landscape almost unique among the landforms of the earth—Meteor Crater. Seen from the air, it closely resembles certain of the craters on the Moon. Situated a short distance south of the highway between Winslow and Flagstaff, Arizona, it is well outside the region of lava flows and cinder cones adjacent to the San Francisco Mountains. This huge hole has a diameter of three quarters of a mile and a depth of six hundred feet below the crest of the surrounding rim which rises a couple of hundred feet above the plateau surface and is composed of abruptly upturned sedimentary rocks. It is now quite certain that this is an *impact crater,* the result of the collision of a large meteorite with the earth at some time between two thousand and twenty thousand years ago. An eight-thousand-ton meteorite about forty feet in diameter, smashing into the earth at ten miles per second, would be adequate to produce such a depression.

Only four or five other impact craters are known. Among them are the Wolf Creek Crater in western Australia, slightly more than a half mile in diameter, and the New Quebec or Chubb Crater in northern Quebec, about two miles in diameter.

A great wall of ice within the crater of lofty Kilimanjaro in Tanganyika, Africa, dwarfs the man on the crater floor. (Annan Photo Features)

WHY VOLCANOES ERUPT

The eruption of a volcano is obviously due to the rise of molten magma to or extremely close to the surface of the earth. If the magma is heavily charged with gas, the eruption is likely to be explosive, analogous to the effervescence of a carbonated beverage when a bottle is uncorked. If there is little or no gas in the magma or if the gas escapes easily because of the magma's low viscosity, the eruption will consist of a relatively quiet outpouring of lava. The difficult questions therefore pertain to the sources and upward movement of the magma, the location of magma chambers, and the nature of the conduits through which the magma approaches the surface.

Recently improved instruments and new geophysical and geochemical techniques have strengthened the belief that magmas originate in the lower part of the earth's crust and outermost part of the earth's mantle, at depths of twenty to sixty miles beneath the surface, by local melting of solid rock. At those depths, rock temperatures everywhere are so great that the material remains in the solid state only because of the immense pressure exerted by the overlying rock. Local melting is due not so much to local heating as to local relief of pressure. The situation is like that of the water in a pressure cooker, which remains liquid even though it is heated far above the "normal" boiling point. Remove the cooker from the stove and open the pressure valve; the water will instantly flash into steam even though it has already begun to cool. Subterranean pressures are locally reduced by certain crustal movements that generally manifest themselves in long narrow zones such as those indicated by the axes of folded mountain ranges or by the great rift valleys. Each of the long and often curving lines of volcanoes on the earth's surface is in a zone of recent crustal movement.

Once the rock changes into liquid in a local magma chamber, it tends to move upward toward the earth's surface. It may melt its way by transfer of heat from magma to chamber roof, or it may rise through fractures in a fissured zone above the magma chamber. All kinds of rock expand when changing from the solid to the liquid state; a chunk of rock will sink when dropped into a cauldron filled with melted rock of the same composition. It is partly in response to the difference in density between molten magma and solid rock that the magma rises through conduits leading upward from a magma chamber, but that movement is generally stimulated also by the strong tendency of the contained gases to expand and thus exert great pressure against the confining roof and walls.

With these characteristics of magmas and magma chambers in mind, it is not difficult to explain the wide variety of volcanic eruptions. Several of the great volcanic cones in various parts of the world are known to be situated at the intersections of major rift or fissure zones. These would be the natural places for the development of pipelike conduits that could feed central volcanoes for a long time. Elsewhere, the permeability of a single fissure zone from below may vary greatly from place to place and one or more volcanic vents may consequently be located along it. Or again, the characteristics of a rift or fissure zone may make possible extensive flows of highly fluid lava with little or no eruption from central vents.

Precise measurements of slight movements of the ground near the top or on the higher flanks of certain volcanoes, notably Kilauea in Hawaii and Usu in Japan, have revealed interesting details concerning the subterranean movement of molten magma. Before an eruption, the ground surface tilts upward toward the summit of the dome-shaped volcano; the whole mountain seems to swell. Following the eruption, the ground settles back and the mountain shrinks. Such behavior, when detected by sensitive tiltmeters located on the flanks of a volcano, provides a trustworthy basis for prediction of an imminent eruption. At Kilauea, the analysis of the pattern of ground tilting indicates a small magma reservoir within the volcano itself

Above: The two sharp-crested ridges stretching to the right of Shiprock in New Mexico are dikes of resistant diabase radiating from the neck of an almost completely destroyed volcano. (Spence Air Photo)

Above, left: Volcanic ash around Elegante Crater in Pinacate lava field, Sonora, Mexico, is interrupted (in left foreground) by a fresh lava flow. (Tad Nichols: Western Ways Features)

or immediately beneath its base, at a depth of only two to four miles below the present ground surface. This reservoir is supplied with magma rising through a conduit from the deeper zone where it originates. When the reservoir is filled with liquid under great pressure from below, the volcano actually swells—the volcanologists say it tumesces or inflates—until the lava erupts and the reservoir is emptied, whereupon the summit region subsides a few inches or feet. In these days of new instruments and techniques, volcanology is truly an exciting science.

HARNESSING VOLCANIC ENERGY

Any practical-minded observer of a volcanic eruption is likely to think what a boon it would be to modern, energy-hungry, industrialized man if some of that wild energy could be harnessed. Actually there is no such possibility in the foreseeable future. But certain of the natural phenomena associated with volcanic activity have already been put at the service of mankind. Thus the famous baths of the Roman Empire derived their water from hot springs in the volcanic region of Italy. And today nearly all the buildings in Reykjavik, Iceland, are heated with hot water from the municipally-owned water system supplied from bore-holes and hot springs in that volcanic island. In both of these places, as well as in many others, the hot water is in part of magmatic origin—the condensation of the water vapor and other gases in a solidifying magma—but in much larger part it is ground water infiltrated from the surface and heated by contact with the warm volcanic rock.

The most successful attempt to use the heat made available by volcanic activity is at Larderello in Tuscany, Italy. Here many natural steam vents are fed by superheated steam that rises under great pressure, through fissures in the rocks overlying a body of magma or recently consolidated igneous rocks at a shallow depth. Several steam wells have been drilled to depths between five hundred and fifteen hundred feet and one to a depth of about five thousand feet. The turbo-generators driven by the natural steam are now supplying about a third of all the thermoelectricity produced in Italy. The steam contains a considerable percentage of boric acid and other gases that testify to its magmatic origin; these are removed in a chemical plant before it is fed to the turbines.

Somewhat similar power plants have been installed in the Wairakei area on the North Island of New Zealand. Its output of electrical energy is at present much less than that at Larderello. Steam wells have also been drilled near Hveragerdi, Iceland, and a natural steam power plant is scheduled for completion there in 1963. The first commercial geothermal power plant in the United States was installed in 1960 in an area of hot springs and fumaroles in Sonoma County, California, known as "The Geysers," although no geyser is located there. The steam from the wells, drilled to depths of several hundred feet, is largely of magmatic origin and rises toward the surface in one of the many fault zones in that state. Thus far, the power capacity of the development is slight, but it will be increased.

13

When the Earth Trembles

Like volcanoes, earthquakes represent the earth at its most cataclysmic, and they sometimes seem even more frightful because they strike invisibly and often without the slightest warning. Suddenly the earth, symbol of all that is solid, trembles beneath one's feet, walls crack and, if it is a major quake, buildings crumble and cities fall into ruins.

At least twenty or thirty earthquakes of great magnitude disturb parts of the earth's surface every year. Any one of them can cause a great catastrophe if it starts near a large center of population. In addition, more than a thousand earthquakes each year are large enough to be recorded on instruments several thousand miles away. Most would be felt by people living near the starting points. If to these are added the earth tremors actually noticed by a few persons within an area of only a few square miles, or recorded at only a few of the thousand or more earthquake-recording stations of the world network, the number of earthquakes per year certainly exceeds fifty thousand. One wonders whether we are justified in referring to land as "terra firma."

Fortunately, most earthquakes of great magnitude originate beneath the sea, scores of miles from land, or in sparsely populated mountains. Hence, the great disasters caused by the quaking of the earth are relatively few. Even so, disasters have occurred somewhere during the last three centuries at the average rate of one every two or three years.

THE MAJOR EARTHQUAKES

One of the most destructive earthquakes in modern history, the Great Lisbon Earthquake, occurred when three violent tremors struck the Portuguese capital on November 1, 1755. Most of the buildings crashed to rubble and more than sixty thousand people were killed. The disaster wrecked buildings in scores of towns in Portugal, Spain and North Africa. Thousands perished as far away as Fez and Mequinez in Morocco, four hundred miles from Lisbon. The shocks were felt throughout much of Europe, probably over an area greater than that of any other earthquake in recorded history. Actually, these were probably a series of related quakes that may have originated hundreds of miles apart. With modern instruments they might have been identified as such rather than as aftershocks of the disturbance at Lisbon.

Earthquakes of similar magnitude struck the lower Mississippi Valley in

1811 and 1812. Because the first great shock, on December 16, 1811, was centered near New Madrid, Missouri, that frontier trading post gave its name to the series of tremors that followed during the next fifteen months. At least two equalled the violence of the first. The loose alluvial soils of the Mississippi bottomlands cracked open, river banks caved in, trees crashed and log cabins collapsed. Large tracts of land rose up several feet, draining swamps; other areas sank and were flooded. A part of the river's floodplain, nearly one hundred fifty miles long and thirty-five miles wide, dropped three to ten feet, and promptly turned into several new lakes, of which Reelfoot Lake in northwest Tennessee is the largest. Had the New Madrid earthquake come a century later, it would have brought an appalling loss of life and property, but in 1811 the few terrified pioneers largely escaped real harm.

Similar changes of level of the ground accompanied the Kutch earthquake of 1819 that centered near the mouth of the Indus River in what is now Pakistan. The sea inundated an extensive area. To the north another part of the Indus floodplain rose as much as twenty feet to form the "Allah Bund" or "Dam of Allah," stretching fifty miles across the lowland. The Wellington, New Zealand, earthquake in 1855 was likewise accompanied by tilting and uplift of the land.

Land levels remained unchanged during the great Charleston, South Carolina, quake of August 31, 1886, although it was felt throughout the eastern half of the United States. In Charleston, scores of houses were demolished and scarcely a building escaped damage. Twenty years later, on April 18, 1906, a major earthquake and fire almost destroyed San Francisco, California. An estimated seven hundred people died and $400,000,000 in property was lost.

The San Francisco earthquake especially interests those who study seismic phenomena. ("Seism" is the Greek word for earthquake; seismology is the science dealing with earthquakes; seismographs are instruments to record earthquakes, and seismograms are the records made on such instruments.) One of the major fractures in the outer part of the earth's crust extends northwestward along the floor of the Gulf of California from its mouth at latitude 22° N., comes ashore and appears in the Imperial Valley, continues past San Bernardino, crosses the west part of the Tehachapi Mountains, goes toward the coast west of San Jose, passes through the western outskirts of San Francisco, cuts across Punta Arenas and Cape Mendocina, and has been traced on the Pacific Ocean floor as far to the northwest as latitude 45° N., a total distance of nearly twenty-five hundred miles. It is really a zone of many closely associated fractures or faults, with several branches leading away from the main zone at various places. From San Bernardino northward for nearly six hundred miles, the system is represented by a single, clean-cut break, the San Andreas Fault or Rift.

The segment or block of the earth's outer crust west of this fracture zone has moved northward—relative to the segment east of it—at least fifteen miles since the start of the Pleistocene Epoch a million years ago. Geologists cannot tell how much of this displacement resulted from northward movement of the western block and how much from southward movement of the eastern segment. The movement took place in a long series of jerks, each a sudden displacement of a few inches or feet. These occurred many years apart and at different places along the rift at different times. Whatever the underlying causes, the impetus to move seems to have been operating constantly. Thus by April 18, 1906, the strain of the western block to move northward relative to the eastern block built up to the breaking point. The sudden rupture extended two hundred and seventy miles along the ancient fault from San Jose to Punta Arenas, with a relative displacement of as much as twenty-one feet at a point thirty miles northwest of San Francisco.

All along the line the fences, roads, and water pipes that crossed the San Andreas Fault were sheared off and displaced horizontally by anything from a few inches up to several feet. As one major block of crust ground past the other, vibrations going out through the bedrock started the quaking of the earth. In the heart of downtown San Francisco, those vibrations spelled utter disaster. The loose surface soil, some of it "made land" and "fill," shifted bodily down slopes; car tracks buckled, pavements were shattered and pipelines broken. Buildings were shaken into various stages of collapse; many frame houses were wrenched and distorted, some by the slumping of the subsoil beneath their foundations. But few well-designed and solidly built structures suffered serious damage. It has been estimated that only five per cent of the total property loss was directly attributable to the temblor (a Spanish word, now widely used as a synonym for earthquake). For three days fires raged out of control because of the broken water mains and the disaster was complete.

Seventeen years later, on September 1, 1923, a similar snapping of the earth's outer crust along a similar fault zone produced an even greater disaster in the cities of Tokyo and Yokohama. Although the fracture there is offshore beneath the sea, its presence and the nature of the movement are known by indirect evidence. The western continental segment of the crust moved northward relative to the eastern oceanic segment. The earthquake was about the same magnitude as that in California in 1906 and the densely populated cities were farther from its place of origin, but the flimsy structures of Japanese cities were much more vulnerable than buildings in San Francisco. Upwards of 150,000 people were killed, 100,000 were injured, and more than 500,000 buildings were destroyed, again more by fire than by collapse.

The Italian cities of Messina and Reggio, on opposite sides of the narrow strait between Sicily and the mainland, were hit by a violent earthquake on December 28, 1908. It severely damaged both cities and took one hundred thousand lives. Two great earthquakes centering near Kanau in western China occurred in December, 1920, and in May, 1927, with an estimated loss of life each time of that same round number of one hundred thousand. Another disastrous shock of major proportions hit New Zealand in 1931 in the vicinity of Hawkes Bay.

In 1950 the upper reaches of the Brahmaputra River in Assam suddenly became a scene of monstrous desolation as a result of widespread landslides triggered by another violent quake. Several of the Brahmaputra's tributaries were temporarily dammed by the slides and some of the dams were washed out, causing disastrous floods that overwhelmed the villages downstream. Some of the landslides were so massive that they gave rise to reports of a "mountain that walked." It was the second greatest earthquake catastrophe in the recent history of Assam, exceeded only by the "Great Assam Earthquake" of 1897 which centered in the big bend of the Brahmaputra, farther

downstream, and was of sufficient magnitude to destroy buildings in Calcutta two hundred miles away.

Northern Algeria suffered a major quake in 1954 that turned Orleansville into a shambles; even its solidly constructed buildings were severely damaged, thousands were killed and many more were left homeless. The earthquake that triggered the landslide in the Madison River Canyon, Montana, occurred on August 17, 1959, and centered near the reservoir known as Hebgen Lake. It was not of great magnitude but is of special interest because the faults where the temblor started were displaced vertically rather than horizontally. On one side of the longest fault, about seven miles long, the ground sank as much as twenty feet relative to the upthrown side. The basin of Hebgen Lake tilted, submerging the north shore and causing the water to withdraw from the south shore. The reservoir was thrown into huge oscillations with upsurges every seventeen minutes that continued for nearly twelve hours. Although the initial surges splashed over the dam at the end of the reservoir, it remained intact.

Three notable earthquakes occurred at widely separated localities in 1960—one on February 29 at Agadir, Morocco, where damage was especially severe in the Casbah, the old native quarter, with the number of dead estimated at ten thousand; the second on April 24 at Lar, Iran, where hundreds of poorly constructed buildings on soft alluvium crashed in complete ruin. The third must be rated as one of the greatest in history. It was actually a

The light-colored scar running across this mountainside from lower right toward upper left is the trace of a new fault. Displacement along this fault caused the Hegben Lake earthquake on August 17, 1959. (Bureau of Reclamation)

261

series of temblors that began on May 21 and continued for more than a month along the coast of Chile from Conception to Chilloe, five hundred miles to the south. One of these was close to the maximum magnitude ever recorded; records at seismograph stations showed that the whole earth vibrated like a great bell struck by a clapper. Many were aftershocks from practically the same place of origin, but at least four of great magnitude originated at widely separated points along a major fault similar to the San Andreas Rift, situated on the ocean floor a few miles offshore and probably about seven hundred miles long.

THE SHAKING EARTH

Relatively few earthquakes are known to have been caused by movement along fractures such as the San Andreas Rift, although the temblors thus originated have often been of great magnitude. Some earthquakes are caused by landslides that jar the ground on which they fall, some by the slumping of sediment on submarine slopes, and many by the eruption of volcanoes. But all are vibrations that spread outward through the body of the earth in all directions from their place of origin. The dimensions of these waves and therefore the intensity of the quake are determined by the amount of energy released at their starting point and by the nature of the earth materials through which they pass.

Movement of molten magma in the throat and crater of a volcano, especially if it involves the explosive expansion of gases, inevitably jars the ground nearby. Such vibrations are numerous during eruptions, but *volcanic earthquakes* are generally of only slight or moderate magnitude. Few have caused damage beyond a radius of a dozen miles from the volcano.

It is the so-called *tectonic earthquakes* that cause the great catastrophes. Most of them result from movements of rock bodies within the earth, hence the label "tectonic" which means "pertaining to the structure of the earth's crust." The movement may be along a fault plane like the San Andreas Rift; it may be a displacement at greater depth along a fracture that peters out before reaching the surface; or it may be the slipping of a layer of sedimentary rock over another layer when, for example, such rocks are folded by mountain-making forces.

In any case, the displacement of the rocks sets up a series of sudden jars each time friction is overcome and jerky movement occurs. Each jar may bring an earthquake. Each movement takes place when stresses that have piled up within the crust strain the rocks until their resistance to movement is overcome. Then the rocks bend or slip, temporarily relieving the strain. Usually a long time elapses before a new strain builds up to the breaking point. For this reason most destructive earthquakes at any one place are separated by scores or hundreds of years. This also explains why earthquakes are frequently grouped into brief periods of unusual seismic activity that recur at long intervals. A strain relieved at one place may increase strain at some other locality in the same general region; thus most of the great earthquakes are followed by many aftershocks before stability is attained. Also some of them are preceded by foreshocks.

The place at which a given earthquake originates is its center or *focus.*

Usually the focus is below the earth's surface, and the point on the surface directly above it is known as the *epicenter*.

Now that instruments for the precise recording of earth tremors are in wide use, seismologists can ascertain the depth of the focus of most earthquakes. About four-fifths of all recorded earthquakes start from foci between ten and thirty miles deep. In contrast to these normal or *shallow-focus* earthquakes, a few originate at depths down to as much as 435 miles–the deepest known.

Deep-focus earthquakes originate between the earth's crust and the earth's central core in the part known as the mantle. Their distribution is limited and puzzling. They are fairly common in South America east of the Andes, at depths of three hundred to four hundred miles, with somewhat shallower ones beneath the Andes. In contrast, the shocks occurring along the west coast and offshore are shallow-focus. Beneath the Japanese archipelago, earthquake depths of about two hundred twenty miles are the rule, and similar quakes are relatively numerous in the South Pacific region, including Indonesia and the Philippines. Foci at depths of about one hundred forty miles are common in the vicinity of the Pamir Plateau in the Hindu Kush Mountains of the Himalayan region. It is probable that the Charleston, South Carolina, earthquake of 1886 was deep-focus, but no seismographs were available to measure it.

Often of great magnitude, deep-focus earthquakes may be felt over large areas without doing serious damage. Even their epicenters are so far above their place of origin that they escape the full effect.

Regardless of the depth of the focus, the vibrations are of several different kinds. They include two types of waves that travel through the earth's interior and several others that travel only along its surface. One of the two

Celestine Pool, Yellowstone National Park, after the water stopped flowing as a result of the earthquake on August 17, 1959. (National Park Service)

263

"body waves" is a *compressional wave;* particles in its path are rhythmically pushed ahead and pulled back so that they vibrate to and fro along the line of advance. Of all earthquake tremors, it travels the most rapidly and is usually the first impulse recorded by distant seismographs. It is therefore known as the P wave—P for Preliminary or Primary; it could just as well stand for push and pull. Waves of this type are akin to ordinary sound waves and, like them, can travel through a gas or a liquid as well as through a solid body.

The other "body wave" is a *shear wave.* Particles vibrate at right angles to the path along which it advances. If you stretch a rope taut from a post and give it a sudden shake, a wave will ripple along the rope. The shear wave itself advances along the rope, but the rope particles only move up and down or from side to side. This type of wave can travel only through solid bodies, a fact which made possible the discovery that the earth has a liquid core. At least, we call the core liquid because these waves do not come through it. Considering the great pressures and high temperatures deep within the earth, the state of material there may not be like liquids we know on the surface. Since shear waves travel more slowly than P waves they are usually the second impulse recorded on seismograms. Hence they are known as S waves—S for Secondary, but also for shear or shake.

P and S waves start from the focus of an earthquake at the same time and traverse identical paths through the solid part of the earth's interior but they travel at different speeds. The difference in the time of their arrival at any point provides a way to compute the distance to their point of origin. A measurement of the distance from three or more seismograph stations gives the location of the place of origin.

"Surface waves" are more complex and are the ones that destroy buildings and sometimes cause the ground to undulate visibly. They often constitute a third set of impulses recorded on seismograms and are known collectively as L waves for Later, or for Longer. The length of these waves may be hundreds of feet or even several miles, whereas the lengths of P and S waves are only fractions of an inch or, at most, an inch or two. L waves originate not at the earthquake's focus but at its epicenter when the P and S waves reach the surface there. As they pass along the earth's surface under a building, they strike hammer-like blows against its base and cause it to sway from side to side.

MEASURING AN EARTHQUAKE

Two distinct measurements of earthquakes are now commonly used; intensity and magnitude. Intensity is measured by effects on the earth's surface—changes of level, landslides, cracks in ground—and on man and man-made structures. It can be estimated by anyone anywhere, without instruments or scientific apparatus. The Mercalli scale of earthquake intensity, now used internationally, has twelve ratings, essentially in terms of the man-frightening and wall-smashing effects, as follows:

    I. Recorded by seismographs but not felt by anyone except possibly by experienced observers at rest.

II. Noticed by only a few persons, generally those at rest on the upper floors of multi-story buildings; resembles vibrations from heavy traffic on a distant road.

III. Slight; felt by many persons indoors; suspended objects sway; vibrations like those from a heavy truck passing close by and lasting long enough for their duration to be estimated.

IV. Moderate; felt by almost everyone indoors and by a few outdoors; dishes and bric-a-brac rattle on shelves; movable objects including some furniture disturbed; floors and walls creak.

V. Generally felt and causes some alarm; furniture, including beds, shifted; pictures may fall from walls and dishes from shelves; swinging bells may ring.

VI. Causes considerable alarm, awakening anyone who may be sound asleep; plaster may crack and some of it may fall; pendulum clocks stop; general ringing of bells.

VII. Strong; general alarm and some panic; people run outdoors; movable objects overturn; some walls crack and considerable plaster falls from ceilings; poorly built structures severely damaged.

VIII. Walls of well-built buildings crack; bricks and pieces of tile fall from chimneys; monuments overturn in cemeteries; substantially built structures considerably damaged.

IX. Structures shifted from foundations; buildings generally damaged and some collapse; general panic; noticeable cracks in ground; some underground pipes broken.

X. Most masonry and frame structures destroyed; landslides on steep slopes; railway tracks bent.

XI. Only buildings of reinforced concrete or steel-skeleton construction remain standing; bridges collapse; all underground pipes and utilities out of service.

XII. Maximum catastrophe; damage total.

Isoseismal lines resembling contours are drawn on maps to show the effects of an earthquake. They are usually numbered according to the Mercalli scale, although many of the older maps use the now generally discarded Rossi-Forel scale. (It was on a scale of ten rather than twelve units and its first six classes were much the same as I to VI above.) All such maps show decreasing intensity with distance from the epicenter and some show the profound influence of diverse earth materials and subsoils upon the stability of man-made structures. Thus the intensity of any given earthquake varies from place to place depending upon many factors besides distance from its focus.

In contrast, an earthquake has only one magnitude. It is expressed in amounts related to the energy released at its source and can be determined only by instrumental observation and measurement. Each higher unit on the magnitude scale indicates ten times as much energy as the next lower one. Thus an earthquake of magnitude 5 involves the release of ten thousand times as much energy as an earthquake of magnitude 1. The magnitude of the largest shock in the series of earthquakes in Chile in May and June, 1960, was 8.5; five others in that series were between 7 and 8. The earthquake that registered intensity IX in Lar, Iran, on April 24, 1960, had a magnitude of only 5.5. The Hegben Lake earthquake in the Madison Valley, Montana, on August 17, 1959, displayed intensity X in that area, but its magnitude was only 7.1, not a truly great shock as earthquakes go.

The great majority of the temblors that have startled the citizens of California since 1900 have had magnitudes between 5.5 and 7.5, but these were shallow-focus quakes, due for the most part to readjustments of the crustal blocks along one or other of the rifts associated with the far-flung San Andreas fault zone. If this kind of earthquake has magnitudes 7.0 to 7.5 it may have an intensity of VII or VIII over a small area, as did the Santa Barbara quake of June 29, 1925.

SAFEGUARDS AGAINST DISASTER

Experiences with many earthquakes indicate what may be done to safeguard life and property. Man can do nothing about the location, frequency or magnitude of earthquakes but he can do much to reduce their destructiveness. The basic principles of earthquake-proof or at least earthquake-resistant construction are well known by competent architects and engineers. One is to tie together all parts of a building—walls and floors, joists and supporting columns—so securely that the whole structure vibrates as a unit. Another is to avoid putting heavy masses, such as large tanks full of water, on roofs or upper floors of lightly constructed buildings. Modern concrete and steel construction almost invariably meets these specifications. High buildings are ordinarily designed to withstand the horizontal pressure of the strongest winds, not merely to support floors and roofs against the steady downward pull of gravity. When earthquake vibrations strike the foundations of a building it is the impact of a blow, often equalling gravity in impetus, applied horizontally. Any modern, well-built skyscraper will sway to and fro in response to such impacts, as it does to sudden gusts of a gale, but it is not likely to be harmed. Several of the ancient cathedrals in Europe have stood secure when buildings around them crashed in ruins, because flying buttresses protected the walls against horizontal stresses.

Equally important is the nature of the ground on which the buildings stand. Earthquake vibrations are three to ten times more likely to damage buildings on "made land" or the "fill" used to reclaim marshes or to produce a smooth building site, than to damage those on solid rock. Four different kinds of terrain may be recognized. Buildings with foundations set on solid rock are least endangered; those on compact glacial till, like that in drumlins, or on partially consolidated sediments are in only slightly greater danger. Structures built on the sand and gravel of glaciofluvial deposits or on the soft alluvium of valley bottomlands will be subjected to much more violent shocks during any given earthquake than those on bedrock. And "made land" is worst of all.

The possibility of earthquake damage should be considered in designing buildings that are to stand on either of the two less stable terrains, even in regions that only experience quakes with magnitudes between 5 and 7 three or four times in a century. If possible, the foundations of structures in areas of "made land" should go down through the fill to the firmer subsoil beneath. The choice of building materials—wood, brick, stone, or concrete—is not

266

In a citrus grove near El Centro, California, rows of trees were offset (at center) by movements along a southern branch of the San Andreas fault. (Charles W. Herbert: Western Ways Features)

important; properly used, one is as safe as another. Even buildings with only one or two floors should be constructed so that walls are securely cross-braced and firmly tied to floors and roofs.

Intense and unreasoning fear commonly seizes people undergoing their first earthquake, however slight. Panic is a grave danger when crowds are gathered in department stores, churches, theaters or schools. Rushing wildly for the street is the most foolish thing possible. The safest place during most earthquakes is a doorway between rooms in the interior of a building, or an inner corner of a room. If the outer walls collapse, wait until the rubble has stopped falling outside and then—if you can—make your way to some wide-open space before an aftershock causes complete collapse of buildings weakened by the main shock.

EARTHQUAKE EXPECTANCY

There is at present no scientific basis for predictions of the exact place, time, and magnitude of future earthquakes. It is not yet possible to measure the accumulating strains within the earth so as to discover in advance the date and locality at which they will be eased by movements of crustal segments.

On the other hand, it is possible to determine with a fair degree of accuracy the earthquake expectancy or hazard for any region. The distinction between earthquake prediction and earthquake expectancy is important.

Life insurance companies can tell how long an average man may expect to live; they cannot predict when any individual will die. The earthquake expectancy in any locality is judged first by the record of past earthquakes in the general region. Then the geologic structure is investigated to find the nature and intensity of the strains accumulating within the earth, and where those strains are most likely to be relieved. From all this information, the *seismicity* (liability to earthquakes) of the area is evaluated.

Any seismologist can predict, for example, that there will be an earthquake in Japan during some specified month in the future, and his prediction will come true; earthquakes of sufficient intensity to damage property have occurred somewhere in Japan on an average of two per month during the last few centuries. Such a prediction, of course, has no value; to be useful, the precise locality and date of the disturbance would have to be foretold.

Regions of high seismicity are closely associated with regions of active or recently active volcanoes. The circumpacific "ring of fire" is matched, for example, by the circumpacific belt of earthquake-prone areas. Volcanic eruptions cause many volcanic earthquakes there, most of them trivial. The ultimate causes of the far more important tectonic earthquakes, which must be the same as the ultimate causes of volcanic activity, will be considered in the next chapter in relation to the conditions responsible for the making of mountains.

Certain regions within the circumpacific belt display greater seismicity than others. One extends in a great curve from the Kamchatka peninsula through the Japanese islands and the Philippines. Another includes New Zealand and a third follows the coast of South America from Peru to the southern tip of Chile. The fourth extends from northern California through the coastal region of Mexico.

Each of the other belts of volcanic activity—the Indonesian region, the West Indies, and the Mediterranean area—is also a region of high seismicity, but there are other earthquake-prone areas far outside such belts. The most notable of these is in southern Asia where a zone of high seismicity extends from Iran eastward to include the Himalayas and adjacent lands extending far to the north in western China.

Regions of moderate seismicity include the St. Lawrence valley, New England and the Atlantic seaboard (in the United States) as well as the lower Mississippi Valley and the northern Rocky Mountain states. Other areas in this category are in southwestern Europe, eastern Australia, and East Africa. Many earthquakes originate in the vicinity of the Mid-Atlantic Ridge, beneath the floor of the Atlantic Ocean, whereas the central basin of the northern Pacific Ocean, except for the Hawaiian Island region, is almost non-seismic.

In contrast to these regions of high and moderate seismicity, the rest of the earth's crust appears relatively stable. It is best, however, to think of that large part of the earth's surface as displaying low seismicity rather than being completely immune to earthquakes. Historical records show that many high-seismic areas have undergone a spasm of activity followed by a lull of a century or more, but modern instrumental records cover only half a century. Almost every year, earthquakes occur in places far distant from previous epicenters. There are probably few, if any, square miles of the earth's surface that have not been disturbed by earthquake vibrations strong

268

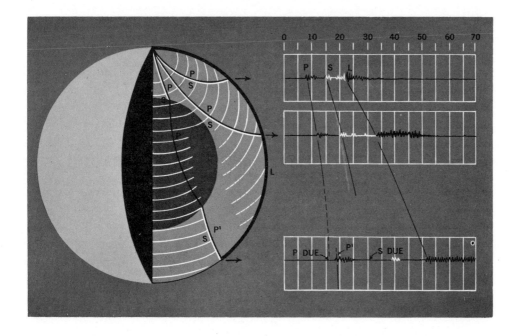

In this cut-away of the earth, the black and white lines diverging from the earthquake center at the top represent travel paths of the P (Push) and S (Shake) waves respectively. The P wave is diffracted by the earth's core but starts a modified S wave in the mantle beyond, whereas the original S wave is blocked by the core. The seismograms at right show the increasing interval between arrival of the P, S and L (Surface) waves at increasing distances from the earthquake center.

enough to be felt or to cause some damage in recent geologic time and that will not again be so disturbed in the next few centuries.

TSUNAMI OR "TIDAL WAVES"

An earthquake of considerable magnitude, with its epicenter beneath the sea or along a coast, often generates movements of ocean water popularly known as "tidal waves." These gigantic waves have no connection with tides; hence it is preferable to designate them by their Japanese name, *tsunami,* as seismologists and geophysicists are now doing. Many earthquake disasters have been compounded by tsunami and on occasion these waves have been more destructive than the quake itself.

The earthquakes caused by the explosion of Krakatoa in 1883, for example, were relatively local affairs, but the tsunami they initiated drowned more than thirty-six thousand persons and caused untold property damage as they swept for hundreds of miles over low-lying coasts of western Java and southern Sumatra. The loss of life and property in Lisbon in 1755 was at least doubled by gigantic waves dashing into that coastal city from the sea a few minutes after its buildings collapsed under the impact of the "great earthquake." Tsunami were also a major factor in the destruction along the coast of Chile in May and June, 1960, and one of them, initiated off the Chilean coast on May 22, caused hundreds of deaths and much property damage the next day far across the Pacific in Hawaii, the Philippines, and Japan.

This type of sea wave is extraordinary. Tsunami travel across the open ocean with a velocity between three hundred and five hundred miles per hour; they measure a hundred to four hundred miles from crest to crest but the wave-height in the deep sea is so small that they have no noticeable effect on ships. Approaching land, they pile up to heights of ten to fifty feet or more and lose speed when they enter shallow water. Along shores, their first apparent motion is a withdrawal of the water as though by an exceptionally low tide, sometimes to a distance of a mile or more. Then in a few minutes, the returning water rushes in at mile-a-minute speed to dash far onshore, overwhelming everything in its path and reaching scores of feet above normal high-tide level.

One tsunami, for example, was generated April 1, 1946, by a severe earthquake on the floor of the North Pacific close to the Aleutian Islands. Four and a half hours later, it reached Honolulu, having traveled 2240 miles at 490 miles per hour. Fortunately the waves were not great and little damage was done there, although it was still able to produce five-foot fluctuations of sea level when it reached Valparaiso, Chile, 8060 miles from its source. The P and S waves from that earthquake traveled to Honolulu in minutes, arriving there four hours ahead of the tsunami.

This provides the basis for the "seismic sea-wave warning system" now in operation in the Pacific region. A network of seismograph stations sends messages to the Honolulu Magnetic Observatory of the United States Coast and Geodetic Survey whenever a severe earthquake is recorded. This permits a hasty location of the epicenter, generally within an hour after its occurrence. If the epicenter is in or near a deep trench on the Pacific Ocean floor— areas considered to be dangerous—a warning is issued, giving the approximate time the tsunami is expected to arrive at various places in the Hawaiian Islands and around the Pacific coast. The system worked well for three destructive tsunami in 1952, 1957, and 1960, but unfortunately only a third of the people evacuated the danger area in Hilo in 1960 and sixty-one persons were killed there despite six hours of advance warning. On at least one occasion many people in a Hawaiian town rushed down to the beach to see the expected great wave. Fortunately it turned out to be a small one or they would all have been drowned. It is difficult to get everybody to take advantage of the knowledge now available to them.

14

The Making of Mountains

The splendor of mountains has always kindled the spirit of man. Artists have painted them in a thousand forms, writers have celebrated them in poetry and prose. Men of religion "lift up their eyes to the hills" and locate their gods on mountaintops; in classical mythology, Olympus was the fabled home of the deities and there they assembled at the court of Zeus; on Mount Sinai, Moses conversed with Jehovah and was given the Ten Commandments; Gautama Buddha attained enlightenment under a bo-tree high in the mountains on the edge of Nepal. Venturesome folk respond to the urge to climb high mountains, no matter how difficult and dangerous the ascent may be; they scale precipitous heights to admire the view and rejoice in "conquering" so great an obstacle.

Geologists, on the other hand, study mountainous landscapes to find out how the rugged topography came to be and to gain an understanding of the forces operating in the earth's interior. This is often more than a quest for knowledge for its own sake; the forces that make mountains are also involved in the formation of valuable metallic ores and fuel resources. Here as elsewhere, science as the search for knowledge goes hand-in-hand with science as the servant of mankind.

The mountain climber rates some mountains as easy to climb, others as difficult, and still others as almost, but never quite, impossible. Similarly, the geologist finds some mountains easy to explain, others more difficult to account for, and still others so baffling that he has not yet found answers to all the questions they raise. Before we are through, our enquiry will therefore take us beyond the established facts into hypothesis and conjecture.

VOLCANIC MOUNTAINS

Among the easiest mountains to explain are the volcanic cones and shield volcanoes. They were obviously built up by eruptions of lava and pyroclastic debris. All of them are geologically recent; any constructed more than a hundred million years ago have long since been eroded away or have left only stumps or the igneous rocks of their magma chambers and the conduits through which they were fed.

Closely related to volcanoes are the mountains produced by the rising bodies of magma that form laccoliths. When such intrusions lift the overlying rocks and arch the surface, a more or less circular area or domelike blister

271

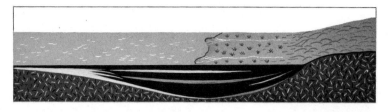

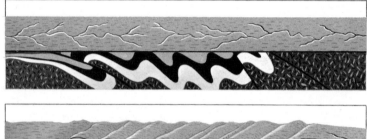

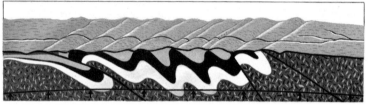

From top to bottom these diagrams show: (1) the Appalachian geosyncline or "trough" filled with sediments largely from highlands to the east (right); (2) the geosyncline compressed by westward movement of the high-lands; (3) the result of erosion that levelled the original Appalachian Mountains; (4) the region as it is today.

stands high above the surrounding landscape. This is soon carved into a group of peaks and ridges by the vigorously eroding streams flowing from its summit. The Henry Mountains in Utah are the classic examples, but the western United States has many other isolated laccolithic mountains.

In certain places explosive eruptions of many closely grouped volcanoes have built vast plateaus of tuff, agglomerate and lava flows thousands of feet above the former land surface. Deeply eroded by running water sometimes assisted by Pleistocene glaciers, such plateaus are now rugged mountains. Parts of the western Andes, the Cordillera Occidental in Peru and Bolivia, as well as the San Juan Mountains in Colorado owe their origin largely to this type of volcanic activity, although in each of these ranges crustal move-ments have increased the altitude.

MOUNTAINS FROM UPLIFTED PLATEAUS

Many of the earth's great tablelands consist of essentially horizontal beds of sedimentary rock, and some have been so deeply and intricately eroded that they are mountains rather than plateaus. Layers of sandstone, limestone and shale containing fossils of marine animals are now thousands of feet above sea level and high above surrounding lowlands, but are still almost as flat as when they were deposited on an ancient sea floor. Evidently, segments of the earth's crust have been lifted bodily upward.

What has happened to the surface of such elevated lands depends upon the length of time since the crustal movement occurred, the resistance to erosion of the rocks, the climate and the river system traversing it before the uplift. As described in Chapter Six, the vast Colorado Plateau in Arizona and adjacent states was uplifted several thousand feet about one or two million years ago. Since then the Colorado River has eroded its spectacular canyon a mile deep, but back from the canyon rim on either side, thousands of square miles of plateau have remained almost unchanged in the nearly arid climate.

Only a few score miles to the north in Utah, the Virgin River and its many tributaries, supplied by more abundant rainfall, have carved the magnificent mountains of Zion National Park in the loftier Markagunt fault block. Many of the mountains there are called "temples" and "thrones"; their detailed sculpture is profoundly influenced by the character of the horizontal beds of sedimentary rock. Even there, however, many almost uneroded remnants of the original plateau lie a few miles from the larger streams.

The Catskill Mountains in New York State are much older. There the sedimentary rocks were uplifted so long ago that nothing remains of the ancient plateau surface. Erosion has not only carved out the valleys between the ridge crests but has lowered the mountain summits far below that ancient surface. The extensive system of essentially horizontal strata ends on the east at a major fault zone along the mountain front overlooking the Hudson Valley.

Similarly the mountainous landscapes in certain parts of the Guiana Highlands in eastern Venezuela and northern Brazil are due to extensive erosion of uplifted flat-lying sedimentary rocks. The famous Table Mountain rising above Cape Town in South Africa is likewise a remnant of an uplifted block of horizontal strata.

DOMAL MOUNTAINS

Vertical movements often produce domelike structures resembling those above laccoliths, but much larger. Erosion of the uplifted rocks may then produce domal mountains, such as the Black Hills in South Dakota. The doming there affected the surface throughout an elliptical area about 125 miles long and 60 miles wide, in the midst of the Great Plains far to the east of the Rocky Mountains. Had the uplift occurred all at once and without concurrent erosion, the dome would have risen 9000 to 10,000 feet above the plain. It took a long time for the crust to warp, however, and erosion began as soon as the surface started to rise. Therefore the dome never reached that height. Today the central peaks rise a little less than 4000 feet above the plains on the east. Harney Peak, 7216 feet above sea level, is the loftiest; Mount Rushmore is much lower.

Erosion of the sedimentary rocks from the summit of the dome has exposed the up-arched basement complex of Precambrian granite and metamorphic rock that make up the central "hills." Those rocks resist erosion more than any others in the region; consequently the valleys are youthful and the slopes precipitous. Surrounding the central, most mountainous area,

the sedimentary rocks crop out in concentric bands, the oldest formation nearest the center and the younger, overlying formations in regular sequence outward from it. The weaker shaly beds underlie broad flat-floored valleys whereas the stronger limestones and sandstones form ridges that curve around the flanks of the dome. Everywhere the layers of rock are inclined downward away from the central hills.

ANTICLINES, SYNCLINES AND OIL POOLS

Any archlike or domal structure or other upward bulge of layers of rock is an *anticline* (Plate 110); similar downward bends are *synclines*. Some are large and with gentle slopes, measuring many miles from side to side; others have steeper dips and may be only a few yards wide. Some are long and narrow, others short and broad. The Black Hills dome, twice as long as wide, is a "doubly-plunging anticline"—an anticlinal fold with the strata dipping downward at both ends.

Smaller doubly-plunging anticlines hold many of the world's valuable oil pools. Sedimentary rocks such as shale and limestone often contain considerable organic material, much of it the cellular tissues of lowly forms of marine life. From this the hydrocarbon compounds of petroleum are generated by chemical reactions, stimulated by the temperature and pressure a few thousand to twenty-five or thirty thousand feet below the earth's surface. The newly generated petroleum may then be squeezed out of these "source beds," as they are further compacted, and move into more porous and permeable rocks such as the layers of sandstone that overlie the shale or are interbedded with it. Thus the spaces between the sandgrains of a "reservoir rock" may be filled with oil and gas. Where such rocks are tilted upward toward the crest of a doubly-plunging anticline, the oil and gas migrate up the dip until they are trapped in the upper part of the fold. This lateral migration within a reservoir rock is caused essentially by the difference in specific gravity between gas, oil and water.

Above: The intricate folding of the layers in these rocks near Yosemite Valley, California, result from intense horizontal compression within the earth's crust. (Andreas Feininger)

Above, left: Sedimentary rocks arched in an anticlinal fold, on the Gaspé Peninsula in Quebec. (Benjamin M. Shaub)

274

All this takes place far below the ground water table in the zone of saturation described in Chapter Ten; oil floats on water at these depths just as surely as anywhere else. The reservoir beds must be so far below the surface that erosion has not stripped away the confining beds above them. In this underground dome natural gas occupies the summit, oil is high on the flanks, and water fills the reservoir rocks farther down the slopes. Finding doubly-plunging anticlines in regions of generally flat-lying but slightly flexed sedimentary rocks, such as the Midcontinent Oil Region in the United States and the Baku Oil Region in the Soviet Union, has frequently been the first step toward the discovery of rich oil pools.

Although broad anticlines and synclines with gently inclined *limbs* (the strata at the sides of the folds) may be produced by vertical movements within the earth's crust, closely compressed folds with steeply inclined or even overturned limbs could hardly be formed that way. They are obviously the result of horizontal compression of the earth's crust (Plate 111). Structures of this type are common in most of the world's major mountain ranges.

FOLDED AND OVERTHRUST MOUNTAINS

Such extensive chains as the Alps and the Carpathians in Europe, the Altai Mountains and the Himalayas in Asia, the Atlas Mountains and the Cape Ranges in Africa, the Appalachians and the Rockies in North America, the Andes and the Serra do Mar in South America, the Great Dividing Range and the Southern Alps in Australia and New Zealand, and presumably the Queen Maud and Pensacola Mountains in Antarctica are all in the category of folded and overthrust mountains. In the making of great mountain ranges such as these, horizontal movements within the earth's crust have far outweighed the vertical movements. The amount of lateral compression varies greatly from range to range, much less, for example, in the Appalachians of Pennsylvania, Maryland, West Virginia and Virginia than in the Alps of Switzerland, Italy and Austria.

THE APPALACHIANS

The history of the Appalachian Mountains begins half a billion years ago at the start of the Paleozoic Era. North America at that time differed drastically from the modern continent. A shallow trough, several hundred miles wide, extended for thousands of miles from Newfoundland to Alabama. On its eastern side, today occupied by the Piedmont Plateau and the Atlantic Coastal Plain, lay a rugged highland. Sediments eroded from it were deposited in the interior sea that invaded the trough from time to time throughout the long succession of Paleozoic periods. Today these sediments are the layers of sandstone and shale seen on many steep slopes of the Appalachian ridges. Interbedded with them are limestones composed of the calcareous secretions and other remains of the marine organisms living in the interior sea.

As these sedimentary materials accumulated in the trough, its floor was periodically warped downward. At the same time, where erosion was wear-

An aerial view across the densely wooded Appalachian Mountains in eastern Tennessee and adjacent states shows how similar in height are the smooth-crested, parallel ridges. The only interruptions in the ridges are occasional water gaps. (William A. Garnett)

ing down the highlands to the east, vertical uplift maintained the highland well above sea level despite the erosion. In parts of the Appalachian region the layers of Paleozoic sedimentary rocks piled one on another to a total thickness of more than thirty thousand feet, yet none of those layers was deposited in water more than a few hundred feet deep.

Any long and relatively narrow region that has received sediments in this way is a *geosyncline*. The Appalachian geosyncline is typical. Nearly every folded and overthrust mountain system, the world around, was once a geosyncline. Several geosynclines are receiving sediments today; the valley of the Po River in Italy, bordered by the Alps along its northern margin, and the valley of the Ganges in India with similar relations to the Himalayas, are geosynclines filled to sea level and above with deposited materials.

Transformation of the Appalachian geosyncline into the Appalachian Mountains began during the Mississippian Period, continued throughout the Pennsylvanian Period and was completed, so far as the rock structures are concerned, in the Permian Period more than two hundred million years ago. This second major episode in the history of this region involved horizontal movements of the earth's crust. The sector east of the geosyncline moved inexorably westward and crumpled its strata into a series of steep-sided anticlinal and synclinal folds, some of them broken by faults and many of

276

them plunging at their ends. Forces about which geologists can only speculate squeezed the thick body of sedimentary rocks in the down-warped trough between the crystalline rocks of the continental border and similar rocks only a few thousand feet below the surface of the continental interior. The rocks in the trough yielded, sometimes like putty, sometimes like a flexible solid, to this horizontal compression. If the folded strata in Pennsylvania were straightened out again, they would occupy a space some twenty-five to thirty per cent wider than at present. Farther south in Tennessee and the Carolinas, the crustal shortening or squeezing was even greater, breaking apart thick wedges of rock, not simply bending them, and shoving them laterally far along the nearly flat surfaces of overthrust fault planes.

This second episode gave birth to the first generation of Appalachian Mountains. After adjustment to the stresses of horizontal movements, a chain of lofty mountains stood where the geosyncline had been. No one knows how high their summits were. An attempt to reconstruct in imagination the now-eroded folds does not give the answer, for the crest of a newly forming anticline began to erode just as soon as it had risen appreciably above the adjacent lands. Thus, the third episode undoubtedly began while the second was still going on.

During that third phase of Appalachian history, the first generation moun-

The sedimentary rocks of La Pierre qu'Abotse in the Vaudois Alps have been tilted up almost vertically. (Photo Klopfenstein)

tains were almost completely worn away. Before the end of the Mesozoic Era the entire region was reduced by erosion to a peneplain with many monadnocks, usually a few hundred feet high, where the most resistant rocks delayed erosion. These were most notable in the southern Blue Ridge and Great Smoky Mountain areas where the ancient, extremely hard crystalline rocks east of the geosyncline had been thrust westward and upward during the horizontal compression.

In the fourth phase, the mountains attained their present form. Beginning in the Cretaceous Period, about a hundred million years ago, and continuing intermittently during the Cenozoic Era, vertical movements caused the former geosyncline to arch upward so that the land along its axis finally stood two or three thousand feet above the land on either side. The movement rejuvenated the streams and permitted them to excavate the modern valleys in the weaker rocks such as the shales, thin-bedded sandstones, and soluble limestones. The stronger rocks—especially the thick-bedded, firmly cemented sandstones and conglomerates—remained standing in the ridges. On either side of the Shenandoah Valley, for example, the even crests of roughly parallel ridges are remnants of the uplifted peneplain. Where master streams like the Susquehanna, the Delaware, and the Potomac, which were traversing the peneplain before its uplift, crossed a belt of resistant rock, they were often able to maintain their courses in the new cycle of erosion by cutting "water gaps" through the ridges while their tributaries were removing the weaker rocks between them.

This sequence of events—sedimentation in a sinking geosyncline, horizontal compression to form folded and overthrust mountains, extensive erosion, upward warping that permits the sculpturing of a second generation

278

of mountains—has been responsible for the present topography of many other "old, worn-down mountains" throughout the world. The Cape Ranges of South Africa closely resemble the Appalachians in structure and topography; somewhat similar are the Serra do Mar and associated ranges in Brazil, as well as portions of the Great Dividing Range in Australia. The hills of Brittany and Normandy and the Vosges Mountains in France, the Schwarzwald in Germany, and the Urals in the Soviet Union display the same pattern.

The folding and faulting of the Appalachian geosyncline had economic consequences of great value. The sedimentary rocks of Pennsylvanian age in eastern North America include seams of coal unsurpassed in thickness, purity and extent. Most of these are in the flat-lying strata of the Allegheny Plateau west of the folded Appalachians, but the vast swamps in which the plant tissues accumulated—later to become coal—extended eastward into that part of the geosyncline now within the state of Pennsylvania. Here the coal seams were folded along with the overlying and underlying beds of clay, shale and sandstone into anticlines and synclines. The compression drove off the volatile hydrocarbons and transformed the bituminous coal into anthracite. Erosion during the third phase of Appalachian history removed the greater part of those anthracite seams but enough remained in the synclinal folds near Scranton, Pennsylvania, to make that region a major source of anthracite coal.

Offsetting this favorable result of horizontal compression is the fact that the pressures and temperatures needed to transform bituminous coal into anthracite are also able to drive out of the rocks any petroleum or natural gas they might contain. Many of the sedimentary rocks in the Appalachians have counterparts in the plateaus and beneath the plains where oil and gas pools have been tapped, as for example, in western Pennsylvania, West Virginia, eastern Ohio, Kentucky and Tennessee; but there are no oil fields in the closely folded mountains.

On the credit side again, the commercial slate and marble in the Appalachian region, from Vermont through Pennsylvania to Tennessee, is a result of metamorphism of sedimentary rocks in the geosyncline during the horizontal compression.

THE ALPS

Like the Appalachians, the Alps are on the site of a great geosyncline, but the strata were deposited in that geosyncline during the Mesozoic and early Cenozoic eras, rather than the Paleozoic. Furthermore most of the folding and overthrusting that transformed it into a mountain range took place in Oligocene and early Miocene time, fifteen to forty million years ago. In such a "short" time, erosion has made no more than a good start in the phase of Alpine history equivalent to the third phase of Appalachian development. The Alps are truly young, rugged mountains.

More important, the Alpine geosyncline was squeezed by much greater horizontal compression than the Appalachian trough. Today, northern Italy is more than a hundred miles nearer northern Switzerland than it was in Eocene time! The prodigious horizontal forces compressed great anticlinal folds, overturned them toward the north, and shoved them forward, one overlapping another. In places the sheared-off limbs of those folds lie nearly flat; for such great sheets of rock, the Swiss geologists use the term *nappes,* the Germans *decken.* On the precipitous face of certain peaks, towering above such resort towns as St. Maurice and Lauterbrunnen, one may see the lower limb of an overturned fold with the rock layers upside down, and higher on the cliff the same beds are repeated in the upper limb, right side up. Some of the *nappes* can be traced southward to the cores of the anticlinal folds from which the sedimentary mantles were almost completely severed. There it is seen that even the basement rocks, the original floor of the geosynclinal trough, also took part in the movement. Granitic rocks were changed to gneiss and ancient sedimentary rocks to schist; Mont Blanc is carved in one such highly sheared anticlinal core and the Matterhorn in another. The whole effect is like that of sliding a thick cloth over a table and crumpling it into several flat folds that overlap like shingles on a roof.

The Alps are in the midst of a line of young, rugged mountain chains strung out across the face of the earth for more than five thousand miles. The chain begins at the west with the Atlas Mountains in Morocco and Algeria, and continues across northern Tunisia before disappearing beneath the Mediterranean Sea. The ancient geosyncline, now represented by closely folded sedimentary rocks, bends sharply toward the north to reappear in Sardinia and Corsica and reach the European mainland in southeast France. There the Alpes-Maritimes rear aloft in a great arc, curving northeastward to join the Swiss and Austrian Alps as the mountain system swings eastward to cross Yugoslavia and the other Balkan countries and continue in the towering ranges of the Caucasus. Crossing southern Asia, it includes the festooned ridges of the Hindu Kush and Himalayas and extends southeastward into Malaya. Some parts of this far-flung chain show more crustal shortening than others, but all of it was a geosyncline throughout the Mesozoic Era and earlier part of the Cenozoic Era. The maximum yield to horizontal compression in the Atlas Mountains at the west seems to have occurred in late Eocene and early Oligocene time, in the Alps during late Oligocene and early Miocene, and in the Himalayas in late Miocene and early Pliocene. Basement rocks were squeezed upward in the Himalayas and the Swiss Alps to a much greater extent than elsewhere.

105. Towering cliffs of dolomite in Italy's Gran Sasso bear testimony to the power of the mountain-making forces within the earth's crust. (Gerhard Klammet)

THE ROCKY MOUNTAINS, THE CASCADES AND THE COAST RANGES

A long, relatively narrow system of young, rugged mountains begins in Alaska as the Brooks Range, curves southeastward to include the Canadian Rockies in Yukon Territory, British Columbia and Alberta, and the Northern Rockies in eastern Washington, Idaho, Montana, and northern Wyoming, and continues as the Southern Rockies from southern Wyoming across Colorado into New Mexico. In its geosyncline sedimentary rocks were deposited throughout much of the Paleozoic and Mesozoic Eras. Horizontal compression, which began at the close of that period, was largely completed about fifty million years ago during Eocene time. The thrust was from the Pacific border toward the continental interior and the crust was shortened

106. The spires and craggy walls in the Garden of the Gods near
Colorado Springs, Colorado, were sculptured from steeply-inclined
beds of sandstone in the eastern limb of the great anticlinal
fold that once arched over the Front Range of the Colorado Rockies.
(Mary S. Shaub)

107. The sedimentary rocks forming the long, jagged ridge known
as Echo Cliffs (above, right) in northern Arizona are tilted downward
toward the left. Some of the layers are more resistant to erosion
than others. (Tad Nichols: Western Ways Features)

108. The horizontal beds of sandstone in White Canyon, Utah (right), are offset by a steeply-inclined fault plane, along which movement has raised the segment at the left a dozen feet or more above that at the right. (Andreas Feininger)

109. Mt. McKinley, Alaska, the highest peak in North America (20,320 feet), is seen here across a pond in a glacial moraine east of Wonder Lake. (Bradford Washburn)

111. The distortion of the limestone beds (bottom) in Lulworth Cove, Dorset, England, resulted from horizontal compression of this part of the earth's crust. (S. Bayliss-Smith: Coleman and Hayward)

110. The up-arched beds of sedimentary rock near Mexican Hat are part of an anticlinal fold that extends for many miles across southeastern Utah. (Paul Miller)

Man and the Earth

112. The terracing of hillsides for rice paddies, like these in Indonesia, demonstrates man's effectiveness in changing the face of the earth. (Harrison Forman)

113. The Corinth Canal, Greece, is one of the oldest of the world's major man-made waterways. (Gerhard Klammet)

much less than in the Alps. The rocks did not respond like putty to produce *nappes,* but like a somewhat flexible and sometimes brittle solid to produce anticlines and synclines, upthrust wedges and overthrust sheets.

In Montana, Paleozoic strata were shoved eastward at least fifteen miles, coming to rest on crumpled Cretaceous shale. In Colorado, the Front Range is essentially a gigantic anticline (Plate 106) with much of the sedimentary rock stripped away to expose the Precambrian basement granite, gneiss and schist. These compose such lofty summits as Pikes Peak and Longs Peak. Along the eastern face of that range, the steeply tilted beds of resistant sandstone form hogback ridges and "flatirons" that are erosion remnants of the limb or side of the great anticlinal fold.

The Rocky Mountain geosyncline was squeezed and crumpled much longer ago than the Alpine geosyncline. Because erosion has therefore been much greater, the Rockies do not stand so high above their surroundings as the Alps and the Himalayas. Even so, the rugged topography, perennial snowfields, and glaciers cause parts of the system to be called "The Switzerland of America." There is, moreover, considerable evidence that many of the ranges in the Rockies were warped upward a few thousand feet in late Tertiary and Pleistocene time.

Overlooking the Pacific Ocean, a third great North American system of folded mountains curves around the Gulf of Alaska from the Aleutian Peninsula. It includes Mt. McKinley (Plate 109) and the St. Elias and Coast Mountains of the Alaska panhandle and British Columbia. Farther southward it embraces the Coast and Cascade Ranges in Washington, Oregon and California. This chain of rugged mountains occupies the site of the Pacific Border geosyncline, parts of which continued to receive sediments as late as Miocene time. Horizontal compression there reached its maximum in late Miocene and early Pliocene time. These are the youngest of the folded mountains in North America, comparable in age to the Himalayas.

BLOCK MOUNTAINS AND RIFT VALLEYS

A wholly different kind of mountain is found in such places as Mongolia, Libya, southern Algeria, and the Great Basin region of Utah, Nevada and California. The mountains in the latter area, which have been most intensively studied, are separated from each other by desert plains and basins, and differ notably in structure and early history. Each was formed from a block of the earth's crust, lifted or tilted upward by vertical movements and bounded on one or more sides by a vertical or steep fault plane (Plate 108).

On the east is the Wasatch Range, hewn from a long and relatively narrow crustal block tilted upward along its western side where the "Wasatch Front" extends in an unbroken line for 180 miles from north to south, past Ogden, Salt Lake City, and Provo, Utah. The "Front" is the somewhat eroded upper part of the west face of the tilted block. At its foot is the trace of a nearly vertical fault plane along which the upward movement took place, most of it in later Tertiary time.

Far away on the west side of the Great Basin region is the Sierra Nevada, probably the largest and certainly the most magnificently sculptured block-mountain range anywhere on earth. There a segment of the earth's crust,

114. Miners at work a thousand feet underground in the Jefferson Island salt mine (left) near New Iberia, Louisiana. (Hubert Lowman)

115. An "open pit" copper mine (below, left) at Ray, Arizona, demonstrates the exploitation of low-grade ore by excavating at the surface. (Charles W. Herbert: Western Ways Features)

116. A "strip mine" (below, right) near Gillette, Wyoming, reaches coal seams not too far below the surface by stripping away overlying rock. (James R. Simon)

nearly four hundred miles long from north to south, rose up during late Tertiary time, but tilted in the opposite direction from the Wasatch. Thus the great wall above Lone Pine and Bishop, California, is the eroded upper part of the east face of the Sierra Nevada fault block. Far back in Jurassic time, huge granitic batholiths had been emplaced in the earth's outer crust there. Erosion during Cretaceous and early Tertiary time stripped away the rocks that had formed the roofs of the magma chambers and exposed the granitic rocks at the surface of the ground long before the crust was shattered and its segments displaced. Thus the High Sierra, near the east side of the tilted block, is essentially a range of granitic mountains eroded by running water and glaciers, where the vertical uplift permitted the carving of deep valleys. Much of the Great Valley of California is floored with the sediments derived from those valleys and deposited on the western, down-tilted surface of the same crustal segment.

Between the escarpment that faces east along the side of the Sierra Nevada and the Wasatch Front which faces west, are hundreds of other block mountains, repeating on a smaller scale this pattern of displaced segments of the shattered crust. Some are tilted blocks with a fault-line scarp on one side and with the other side buried beneath piedmont alluvial deposits. Some are separated from their neighbors by rift valleys, produced where a long narrow block dropped down between adjacent blocks that were hoisted upward on each side. Death Valley is the longest and deepest of the rift valleys in the Great Basin; the roughly parallel fault-line scarps along its walls are easily recognized as eroded faces of uplifted blocks even though no trace remains of the original fault-plane surfaces.

The different vertical movements responsible for block mountains and the valleys between them are in striking contrast to the horizontal compression that produces folded and overthrust mountains. The latter raises the surface vertically, either along the crest of a compressed anticline or in general by its thickening of the sediments when their width is reduced, but it is obvious that no lateral compression is involved in the making of block mountains. On the contrary, there is good reason to believe that some lateral stretching of the crust was necessary to permit the tilting of the blocks and the dropping of the segments beneath the rift valleys.

This idea is further strengthened by an examination of such truly great rift valleys as those in Asia Minor and Africa, in comparison with which even Death Valley is puny. The rift valleys there are steep-sided depressions, hundreds or thousands of miles long and a few tens of miles wide. Each was formed when a wedge-shaped block of the earth's crust dropped. Among them are the trough occupied by Lake Tiberias (Sea of Galilee), the Jordan River, the Dead Sea, and the Gulf of Aqaba, and the even larger depression filled nearly from side to side by the Red Sea. In East Africa the complex of rift valleys includes the spectacular depression occupied by the string of lakes from Albert in Uganda, through Edward and Kiva to Tanganyika, the bottom of which is nearly a half mile below sea level. To produce structural depressions like these, the outer crust must have ruptured and spread apart in horizontal movements so that the valley segments could move downward.

The survey we have just made of the many kinds of mountains makes it evident that in their origin diastrophism (distortion of the earth's crust resulting from internal forces), vulcanism, and erosion are all involved. Of

An aerial photo of long hogbacks and intervening barren valleys north of Old Henderson, Nevada, shows sedimentary rocks dipping down toward the left as a result of mountain-making movements within the earth's crust. (William A. Garnett)

290

those processes of geologic change, diastrophism is by far the most important; without it there would be no great mountain chains strung across the face of the earth. It is also the most perplexing of geologic processes, presenting many unsolved problems. We must face the fact that the causes of mountain-making movements are still shrouded in mystery. For that very reason, if for no other, the nature of those movements is one of the most exciting phases of modern geology and is the subject of much research in geophysics.

Movements of the earth's crust may be analyzed in terms of four directions. The dominant component of the movement may be vertical, either upward or downward; or it may be horizontal, either from compression resulting in crustal shortening, or from extension resulting in crustal stretching. Of the two directions of horizontal movement, the compressional have been greater and more widespread than the extensional.

ISOSTATIC ADJUSTMENT

The floors of all the great geosynclines, about which we have pertinent information, moved downward while the sediments were accumulating. In the Appalachians and other mountains, the geosynclinal belt moved upward after the first generation of mountains had been largely eroded away. It looks as though the downward movement was due to the weight of the sedimentary deposits and the upward movement to the removal of mass by erosion. This idea is supported by the record of crustal movement in northeastern North America and northwestern Europe at the close of the Pleistocene ice age. In both localities the portion of the continent that had been covered by thick sheets of ice rose when the ice melted. The weight of the ice had depressed the crust; its removal permitted the rise toward pre-glacial positions. It is like a raft that sinks lower in the water when swimmers climb onto it and rises again when they dive off. Such response of the earth's crust to loading and unloading is the essence of a principle known as *isostasy* (equal standing or balance); the movements are termed *isostatic adjustments*.

Precise measurement of the force of gravity at many thousands of places on land and many hundreds at sea have firmly established the fact that the various segments of the earth's crust are generally in isostatic equilibrium. The major irregularities exist because they are "floating" upon the denser, somewhat plastic interior. Continents stand higher than sea floors because of their lower density. Mountain chains and plateaus are higher than lowland plains because there is a greater thickness of low-density or "lightweight" material beneath them. (An iceberg sticks up because it has so much material under the water while an ice floe does not.) As we have said in Chapter Two, the thickness of the earth's crust and the depth to the moho vary from place to place. Isostasy proclaims that the rocks composing the earth's mantle are potentially plastic at depths of thirty miles or more because of the tremendous pressure at those depths. The forces of diastrophism are operating upon and within a relatively thin shell of normally rigid, brittle, low-density rock, resting or floating upon and supported by a thick shell of quasi-plastic, higher-density rock.

This concept of earth structure is basic to any hypothesis of mountain-making movements. It is essential to the explanation, given above, of the

downward movement of geosynclines while sediments accumulate in them and the up-doming of mountains when their summits are eroded away. It also suggests many other possibilities. For example, on the huge tilted fault block of the Sierra Nevada, the weight is being shifted by erosion of its upper half and by deposition on its lower half; small displacements have occurred in geologically recent time along the fault plane on its eastern side. Presumably these vertical movements were isostatic adjustments. If so, there will be more of them in the future and each will initiate an earthquake.

Isostatic adjustment helps to explain some mountain-making movements but a radically different explanation is needed for the closely crumpled strata, the overturned folds and overthrust structures observable in such linear mountain ranges as the Alps, Himalayas, Andes, and Rockies, where intense horizontal compression is obvious. To understand the nature and causes of this kind of crustal movement has been a basic objective of much research during almost the entire history of the science of geology. Consideration has been given to a great variety of concepts and theories, but only recently has attention been concentrated upon the exciting new ideas clustered around the concept of "plate tectonics."

The "Four Sisters" in the foothills of the Santa Rita Mountains, Arizona, were carved from the limb of a great anticline. (Charles W. Herbert: Western Ways Features)

PLATE TECTONICS AND THE SPREADING OF OCEAN FLOORS

Tectonics is a term long used to designate the study of the broader structural features of the earth and their causes, but the term plate tectonics first appeared in the vocabulary of earth scientists in the nineteen-sixties. It embraces the theory that the earth's crust is a mosaic of a dozen or more gigantic slabs or plates, ranging in thickness from about 70 kilometers (about 45 miles) to about 150 kilometers (about 100 miles) and in area from a few thousand square kilometers to several million square kilometers. For example, the large North American Plate extends all the way from the submerged Mid-Atlantic Ridge to the west coast of North America, whereas the Caribbean Plate, bordering it on the south, has approximately the area only of the Caribbean Sea. Even smaller are the Turkish Plate, with about the same area as Turkey, and the Hellenic Plate, only slightly larger in area than

Greece, both of which are tucked into the mosaic between the large Eurasian Plate toward the north and the African Plate on the south. Each plate has great internal rigidity and is in more or less constant movement in relation to neighboring plates and to the axis of rotation of the earth as a whole.

The boundaries of individual plates are of three types. One of these is best illustrated by the conditions now known to exist on the floor of the Atlantic Ocean where the topography has been mapped in great detail in recent years by means of sonic depth-finders. Here there is a submarine ridge system rising four or five thousand feet or more above the adjacent ocean floor, extending southward in a sinuous line from Iceland, where its summits are above water, through the entire North Atlantic and South Atlantic oceans then turning eastward and finally northeastward, far south of Africa's southern tip, into the Indian Ocean. This Mid-Atlantic Ridge consists essentially of two parallel "rises" separated by a narrow trough or rift valley. Paleomagnetic data indicate that the ocean floor is moving away from the ridge system on either side at velocities on the order of two or three centimeters (about one inch) per year. These movements are referred to as "sea-floor spreading." Plate-tectonic theory attributes this spreading to the upwelling of magma from the outer part of the earth's mantle into the rift zone of the ridge system, and there producing new plate material. The new material makes room for itself by shoving aside the older part of the plate. This type of plate boundary is thus found along the axes of submarine ridges or rises where plates diverge from each other.

A second type of plate boundary is almost the antithesis of the one just described. Where two plates impinge upon one another, as a consequence of lateral movement, the leading edge of one may "dive" or "sink" beneath the edge of the other. The downwarped edge of that plate is destroyed by chemical action or simple melting in the hotter environment it enters. The zone in which such activities take place is known as the "subduction zone," and the boundaries of plates involved are "subduction" or "sink" boundaries. Subduction zones are characterized by extensive mountain-building, great volcanic activity, and numerous strong earthquakes.

A third type of plate boundary is more or less intermediate between the two just described; if the first-mentioned is thought of as positive, because new plate material is generated along it, and the second as negative, because plate material is detached and destroyed along it, this third type might be considered neutral. It is found where impinging plates slide past each other without subduction, but produce a zone of transform faults such as the San Andreas fault, along which the relative displacement of the two sides is predominantly horizontal, with little or no vertical displacement.

Without going into more technical details, it would appear that the concepts of plate tectonics account nicely for such features of the face of the earth as the majestic mountain ranges formed on every continent during the last 150 to 200 million years, the volcanoes in the "ring of fire" surrounding the Pacific Ocean and elsewhere, the great rift valleys of Africa and Asia Minor, and the localities where disastrous earthquakes are most likely to occur. Some aspects of the older theory of continental drift have been revived, and the apparent "fit" of the western border of Europe and Africa with eastern edges of the American continent, like pieces in a jigsaw puzzle,

These diagrams indicate the sequence of overthrusting that resulted in the present structure—shown in the bottom block—of the eastern ranges in the Montana Rockies.

again has significant meaning. There is, however, considerable difficulty in explaining the origin and location of the older mountain systems, such as the Appalachians, the Urals, and the Caledonians, the structural features of which were formed in Paleozoic times, more than 200 million years ago. Serious questions also arise concerning the adequacy of the driving mechanism mentioned above: the shoving aside of diverging plates by newly forming plate material.

CONVECTION CURRENTS IN THE MANTLE

In responding to those questions, it will probably be necessary to bring under the umbrella of plate tectonics certain hypotheses, which also began to develop in the nineteen-sixties, concerning convection currents in the earth's mantle. The central idea here is that, given an abundance of time, great confining pressure, and appropriate distribution of high temperatures, the quasi-plastic material composing the earth's mantle, or at least the outer part of it, would be subjected to thermal convection currents analogous to those in a liquid in a pot on a warm stove. Such currents could form "cells," elongate oval in cross section and with the current moving horizontally below the earth's crust as it completed its appointed multimillenial round between bottom and top of the "cell." Locate "cells" of appropriate dimensions and directions of convective "flow" beneath plates and the driving mechanism necessary for plate movement would be greatly strengthened. The spatial relations between plate and "cell" could well be causal rather than merely coincidental. Unfortunately, although many measurements of heat flow from interior to surface have been made in recent years, both on ocean floors and land masses, not enough data are yet available to support a speculative hypothesis such as this. Pending further research, the fundamental causes of mountain-making movements remain an enigma.

15

Man and the Earth

Of all the living things on earth, whether animal or plant, man is the most effective agent of geologic change.

We have seen, for example, how marine creatures make limestone and how land plants form coal seams, how vegetation aids in weathering rock and how earthworms, bacteria, and other organisms benefit the soil, how plant cover influences the work of wind and running water, how organic acids from vegetation turn up in ground water, and how coral reefs may determine a shoreline. But all these are minor compared to the effect of man.

Man has had this immense effect unwittingly as well as intentionally. In trying to shape his environment to his desires, he has had an effect on the landscape that he did not dream of, and sometimes the geologic consequences have been nothing short of calamitous.

Man is of course not unique in the animal kingdom in changing the face of the earth in significant ways. Beavers build dams across a stream to produce a pool deep enough so that it will not freeze to the bottom in winter and in which their "lodges" will be secure against enemies. Termites, especially those of Africa and Australia, construct "termitaria" as much as twenty feet high. In comparison with man's dam-building and earth-moving, however, all such achievements fade into relative insignificance.

The earliest activities of man as a geologic agent were associated with his need for food. Men of the Old Stone Age, like primitive peoples today, made much use of fire in food-gathering and food-growing. Hunters frequently employed fire-drives to flush out their prey; herdsmen set fire to the brush and undergrowth in forests to extend grazing land; farmers burned away the stubble after harvest as well as the natural vegetation from previously untilled areas to prepare for the next season's planting. The man-made fires had a revolutionary impact on the processes of erosion. On every continent, vast areas hitherto protected by a cover of dense vegetation were exposed to slope wash, flash floods and rapid gullying. The rate at which land was denuded by running water and blowing winds has been far greater during the thousands of years in which mankind has been disturbing the balance of nature than in any similar period in earlier geologic history.

The European record is especially shocking. Until about A.D. 900 more than four-fifths of central Europe was forest-clad; a thousand years later, in A.D. 1900, less than a quarter of it was wooded. Much of the deforestation was due to agrarian expansion in the Middle Ages, but a good part of it was

296

the result of the beginnings of the industrial revolution of the sixteenth and seventeenth centuries. Glassworks and soap factories needed more and more wood ash; production of tin, lead, copper and iron depended upon timber for pit props in mines and charcoal for fuel. Indeed, it was probably the imminent exhaustion of charcoal, more than the superiority of coke for smelting iron ore, that stimulated the mining of coking coal in the eighteenth century and thus initiated the development of the coal industry as the capstone of the industrial revolution of the nineteenth century.

Today all industrialized countries share the economic and cultural problems stemming from vanishing woodlands. Not least is the effect of deforestation on the runoff of surface water. The increasingly disastrous torrents pouring down Swiss mountainsides during the latter half of the nineteenth century had to be countered by extensive engineering works. The gashing of vegetated slopes by gullies in many places throughout the world has swept away vast amounts of topsoil and ruined millions of acres of potential farmlands. In addition, the increased runoff from deforested slopes frequently swells rivers to flood stage. Once man has disturbed the vegetative cover, maintenance of the normal regimen of streams is all but impossible.

TERRACE AGRICULTURE

Terrace agriculture is among the oldest of all techniques for the conservation of soil resources. It has been practiced in the Philippines, Indonesia, Japan, Lebanon, Syria and Peru for thousands of years (Plate 112). The shaping of steeply sloping land into nearly level benches, rising steplike, was doubtless primarily intended to provide additional tillable ground, but it also prevented gullying and retarded slope wash. Many of the terraces on the steep sides of valleys in the Andes were constructed in pre-Columbian times with retaining walls of stone, some still in use today. They resemble terraced vineyards along the Rhine Gorge in Germany and on the southward-facing hillsides overlooking Lac Leman in Switzerland.

On gentler slopes the terraces are wider and the risers are commonly low walls or small ridges of earth protected from erosion by grass or other thick-growing vegetation. This is the way the rice paddies on many hillsides of India, Sumatra, Bali, Japan, and similar warm, well-watered lands are constructed. Even without large earth-moving machinery, men have always managed to shape the landscape so as to permit the harvesting of foodstuffs in far greater abundance than the landforms could naturally supply.

MAN-MADE LAKES

The construction of earthen dams across streams in order to store water also dates back to prehistoric times. Although most of the huge, spectacular dams of modern times are made of concrete the earth embankment is still the most common type. Dams are built to redistribute water in time and space. Some merely divert water, generally for irrigation; rarely is there much of a reservoir above such a diversion dam. Navigation dams make rivers deeper and regulate the flow throughout the seasons for navigation purposes. They ordinarily have locks such as those in the Rhine between Basle and Strasbourg and in the Mississippi upstream from Rock Island, Illinois, through which ships may pass from one level to another. These dams interfere with the natural regimen of rivers and of course modify their work as geologic agents, but they do not greatly alter the landscape.

In contrast, storage dams produce reservoirs or man-made lakes that change the earth's appearance and have far-reaching effects upon the life of the region. The stored water may be used for cities or industries, for irrigation, flood control, river regulation, power production, or recreation.

Man-made lakes in arid or semi-arid regions are like oases in a desert. Lake Mead, Nevada, in which the waters of the Virgin and Colorado Rivers are impounded by the huge Hoover (Boulder) Dam, has revolutionized the lives of the people in that region and has definitely improved the local climate. The new recreational facilities, pleasant home sites, and agricultural and commercial potentialities are nearly as significant for the surrounding region as is the hydroelectricity produced there for such distant metropolitan centers as Los Angeles. In Algeria an even greater amelioration of socio-economic conditions is beginning to appear with the impounding of another vast lake behind the recently completed Iril Emda Dam.

Similarly, the lack of natural lakes in many areas of Europe and North America that lay beyond the reach of Pleistocene glaciers is being rapidly remedied by the construction of storage dams. Most of these are designed for both flood control and power production. In Germany, for example, the man-made Edersee, impounded by the Eder Dam southwest of Kassel, is as lovely as Chiemsee, a natural lake in the glaciated area southeast of Munich, and provides similar recreational facilities. In the United States, the Lake of the Ozarks behind the Bagnell Dam on the Osage River brings to south central Missouri many of the advantages of the glacial lakes in Minnesota and Wisconsin.

As noted in Chapter Six, dams are sometimes built to serve several conflicting purposes. For flood control the reservoir must be partially empty before the flood season, whereas for maximum power development it must be full or nearly full at all times. Large fluctuations of water level reduce the recreational uses as well as the esthetic value of a reservoir. Where water is stored for municipal supply systems, the possibility of pollution precludes recreational facilities. Such considerations lead inevitably to the conclusion that the management of water resources should be planned for an entire drainage basin as a unit, with full consideration of all the consequences of interfering with "the balance of nature." Various kinds of dams at various places should be so coordinated as to bring maximum economic, esthetic and social returns to all the people of the region.

MANAGING THE RIVERS

Such management of river systems is being attempted at only a few places in the world. Not one major river system is yet fully regulated according to modern engineering, administrative and biological techniques. Of the eight basins covering a million or more square miles, only the Mississippi and the Nile have more than minor control works. The closest approach to integrated

management is found on such medium-sized streams as the Rhine in Germany, France and the Netherlands, the Rhone in Switzerland and France, the Tennessee in the United States, and the Volga and the Don in the Soviet Union. The San Joaquin River system in California has also been put under extensive management, but for a single purpose—irrigation. There is a growing need for extensive development of such rivers as the Yangtse, Huang, Mekong, Nile, Niger, Tigris-Euphrates, Danube, São Francisco, and lesser streams in densely settled, underdeveloped areas. The remaining years of the twentieth century will probably be a period of extended control of these rivers, as political and economic conditions permit.

Certainly the greatest and most productive works remain for the future. These will include, among others, important installations on the three largest rivers of all (in terms of average volume of discharge), the Amazon, the La Plata-Parana, and the Congo. The basins of these rivers have sites for water storage and power development several times the capacity of the largest thus far developed anywhere in the world.

Surprisingly, the use of rivers for transportation has increased markedly in recent years despite the extensive development of other transportation facilities. The rivers and canals in France carried fifty per cent more traffic

Terracing of hillsides to improve soil resources has long been practiced in the Calea Valley, Peru. (Aerial Explorations, Inc.)

in the late 1950s than in the years immediately prior to World War II, and those in Belgium about forty per cent more. In the United States, the tonnage of inland waterway traffic in 1962 was nearly twice that of 1947, and the Soviet Union more than doubled such traffic between 1952 and 1961. The increase is in part due to improvement in navigation, the primary objective of which is to stabilize a stream channel at a depth adequate for shipping. Stream banks are faced with a layer of stone to prevent erosion and the shifting of channels; where necessary, "contraction works" are installed to concentrate a wide, shallow, or braided stream in a single channel; low-head dams with locks are installed in order to maintain navigable depths; and loading basins are excavated at "harbors" alongside the main channel. Canals may be constructed to connect one river with another, such as the Volga-Don canal in the Soviet Union, or to bypass unmanageable rapids such as those in the St. Lawrence River.

The integration of all the various objectives and techniques of river management is still in the experimental stage. The ways of nature and the needs of human beings require more study before a complete handbook of principles can be written. The Tennessee Valley Authority has been the closest to the ideal so far, especially because in that "pilot project" an effort was made to combine the management of water resources with measures to improve land use and to relate river engineering to the entire economy of the region. That is why the Indonesians pay it tribute by referring to their work in the Sadany River Valley as "the Little TVA."

RESHAPING THE SEASHORE

From time immemorial, people dwelling on seacoasts where waves and currents batter the shore have tried to protect their dwindling land from the onslaught of the sea. Since ancient times it has been customary to build seawalls and embankments but sooner or later these have generally proved inadequate. At some unknown date in antiquity, it was learned that groins and jetties built at right angles to the beach line were far more effective. Today, thousands of miles of seashore are more or less completely stabilized by groins such as those at the foot of the cliffs on each side of the English Channel, or along the beaches on the east coast of Florida. If groins, jetties and seawalls are promptly repaired after damage by storms, man may succeed in maintaining the stability of a seashore.

In addition to this policy of containment there is the more aggressive one of reclaiming submerged land. The most extensive projects are in the Netherlands, where the Rhine, the Meuse (Maas) and the Scheldt rivers have their deltas in the North Sea. Here a long line of barrier beaches capped by sand dunes has been built up by longshore currents since the withdrawal of the Pleistocene ice sheet. The construction of dikes to protect these lowlands from inundation began as early as A.D. 900. By the end of the nineteenth century, more than a million acres of farmlands, pastures and town sites had been claimed from the sea. This involved the transformation of lagoon areas between the barrier beaches and the higher parts of the deltas into polders, each with enclosing dikes and drainage canals. With the more efficient twentieth-century engineering techniques, reclamation is proceed-

300

Above: Hoover (Boulder) Dam and Lake Mead, Nevada-Arizona. (Spence Air Photo)

Above, right: Roseland Dam, Alpes Maritime, France, not long before it was completed in 1961. (McGraw-Hill Yearbook of Science and Technology)

ing on an even more extensive scale. The Afsluitdijk, completed in 1932, stretches for twenty miles across the mouth of the Zuider Zee from North Holland to Friesland and is a remarkable engineering feat. Systematically the greater part of the former embayment is being transformed into polders which by 1980 will add a million acres to the land area of the Netherlands. Still more ambitious are the Wadden and Delta projects which will close the gaps between the barrier beaches forming the West Friesian Islands, now exposed to the fury of North Sea storms, and seal off the entrances to the estuaries of the Maas and Scheldt rivers. Nowhere else on the margins of the seven seas are shores being reshaped on so vast a scale.

Men are also concerned with seacoasts as places from which to set forth on voyages. Each of the great seafaring nations of antiquity was located on a coast favored by natural harbors. Most of these, however, could shelter only a few small ships, and as soon as larger ships were built and great fleets assembled, it became necessary to enlarge the port facilities. Breakwaters, piers and moles were constructed, for the most part of loose stones or rubble, at Eretria, Alexandria, Tyre and elsewhere along the eastern Mediterranean shore before 1000 B. C. A common practice was to join an offshore island to the mainland by means of a causeway, and shelter ships on the lee side.

These ancient practices have been refined and expanded. Natural harbors adequate for the large passenger liners, oil tankers and freighters of the mid-twentieth century are very scarce; only the harbors at Halifax, Havana, and Rio de Janeiro come to mind and the ship channels in them must be dredged frequently to maintain depth. Nearly all the world's important coastal harbors would be of little use without the breakwaters built during the last two hundred years. At places such as Cherbourg, Cape Town, and Dover, man-made structures supplement the inadequate shelter provided by the natural harbor. Elsewhere, as at Los Angeles, Callao, and Madras, the pristine shore affording no more than an unreliable roadstead, the harbor is entirely the work of man.

Seafarers, knowing that a straight line is the shortest distance between two ports, resent any peninsula that juts far out from land athwart their route. Many an ancient mariner of the eastern Mediterranean must have expressed a wish that Jove had seen fit to extend the Gulf of Corinth a few miles farther to the east, thereby transforming it into a strait and making the Peloponnesus an island instead of an extension of the Balkan peninsula. What a boon that slight modification of the coastline of Greece would have been, shortening his voyage from the ports of Italy to those of Asia Minor by at least two days and permitting him to avoid the dangerous passage around the capes at the southern tip of the peninsula! Responding to this challenge presented by the four-mile-wide Isthmus of Corinth, the ancient Corinthians constructed a shipway using wooden rollers and then charged tolls for dragging small ships across. Later, Julius Caesar, Caligula and Nero, each in turn, started the digging of a canal. The cut that Nero began was actually incorporated in the modern canal when the dream of the ancient navigators was finally realized in 1893 (Plate 113).

Several canals in similar locations in other parts of the world are even more important for modern trade routes. The Kiel Canal across the base of the Jutland peninsula in north Germany is fifty-four miles long and connects the Baltic with the North Sea. The Cape Cod Canal across the narrow base of Cape Cod in eastern Massachusetts connects Cape Cod and Buzzards bays. With dredged approaches it is thirteen miles long and provides a shorter, safer passage than the open-sea route between Boston and New York.

The distribution of the continents imposes two great north-south land barriers to east-west water-borne traffic, the Eurasian-African land mass extending 7600 miles from north to south and the two Americas 8800 miles. The bulk of the world's trade moves east and west in the low and middle latitudes, and only at their southern ends is ocean navigation uninterrupted around the globe. But approximately 4800 miles north of its southern end, each of these long land barriers is narrowed to a width of less than a hundred miles: at Suez in the eastern hemisphere and at Panama in the western. The challenge to breach those two isthmuses has been felt from early times.

Attempts to dig a canal at Suez to connect the Mediterranean and Red seas date back at least to the fifth century B. C., but it was not until 1869 that the eighty-eight-mile canal was first opened to traffic. It has been deepened and widened at various times since then. The results of this severing of one continent from another reach far beyond world trade into the tense area of international politics. The simple fact that the water route from Liverpool to Bombay is forty-six hundred nautical miles shorter by way of Suez than around the Cape of Good Hope has had a subtle but significant impact on the life of every person alive today.

Much the same can be said for the Panama Canal, opened to traffic in 1914 after a long series of political maneuvers, economic failures, and engineering frustrations during the latter half of the nineteenth century. The system of great locks, the huge Gatun Dam, and the Culebra Cut through the continental divide were marvels of engineering. The Panama Canal is a locked canal, not a sea-level one like the Suez; large ships are lifted through locks to the fresh water of Gatun Lake, 85 feet above sea level. Nearly half of the fifty-mile voyage, from deep water in one ocean to deep water in the other, is on the lake. Thus the motto on the official seal of the Panama Canal Zone—"a land divided; the oceans united"—is not strictly true in the geologic

sense. Before the end of this century, however, man may truly separate North and South America by cutting a sea-level canal across the isthmus either in Panama or Nicaragua. The present canal is inadequate for the steadily increasing traffic. Once constructed, a sea-level canal is less expensive to operate and maintain than a locked canal of similar capacity, and its tolls may therefore be lower. Moreover, low-cost nuclear explosives for excavating rock may soon help to overcome the economic obstacles to so vast an enterprise.

The desire to send ships by the most direct routes is matched by the desire to travel overland by the shortest routes. Bridging rivers is as ancient and worthy an enterprise as digging canals, and tunneling mountains as draining swamps. Man can thus change his physical environment almost at will. The Golden Gate bridge at San Francisco, California, has the effect of straightening the coastline so that automobiles can be driven directly along the shore without detouring around the bay. When completed, the causeways, bridges and tunnel across the mouth of Chesapeake Bay will straighten the Atlantic shoreline in Virginia and join the "Eastern Shore" with the rest of the Old Dominion without the use of ferries or the long detour around the head of the bay.

Similarly, men have changed islands into peninsulas by tying them to the adjacent mainland by causeways, bridges or tunnels. Venice, which was once a group of islands off the northern coast of the Adriatic, is the most famous example of this. Denmark's Fyn is in effect no longer an island, by

Irrigation of cotton fields near Eloy, Arizona. (Charles W. Herbert: Western Ways Features)

303

virtue of the Fredericia-Middelfart causeway and bridge. Nor is Manhattan truly an island as far as New Yorkers are concerned; they drive through tunnels beneath the Hudson to New Jersey or beneath the East River to Long Island. In Nova Scotia, Cape Breton Island is today a peninsula, thanks to a massive causeway across the Strait of Canso. The North Frisian island of Sylt has been tied to the mainland of Germany by the "Hindenburg Dike," a causeway topped by railroad tracks; and, farther north, the Danish island of Römö is now joined to neighboring Jutland by a causeway carrying a broad highway. In neither case has nature accepted the change passively; shore currents are rapidly depositing silt and sand to broaden the isthmus of each new peninsula. Alongside the Römö embankment, fertile land is rising from the sea, encouraged by low groins and drainage ditches put in by reclamation experts. Here man has made the forces of nature his allies rather than his enemies.

EXPLOITING MINERAL RESOURCES

The use of mineral resources to make tools, weapons, machinery and consumer goods began when a remote ancestor of man first flaked a piece of flint to a cutting edge or ground grain into meal between two slabs of stone. Slowly at first but with increasing ease and speed, especially in the last few centuries, men have been extracting metallic ores and mineral fuels from the earth's crust to use in the manufacture of countless articles and to produce mechanical energy. As a consequence, the most prominent features of the landscape in many places are man-made. Today's powerful earth-moving machinery—bulldozers, draglines, power shovels, etc.—makes man an agent of erosion and deposition, more effective in such places than the geologic agencies with which this book has been so largely concerned.

Open-pit mines (Plate 115), from which the ores of such metals as iron, copper, aluminum and tin are extracted, rival volcanic craters or calderas in size. Strip-mining for coal (Plate 116) has made ugly scars on many a hillside where the "overburden" above the coal seam has been stripped off and left in piles of rubble. The recovery of gold from river gravel is now accomplished, notably in Alaska, by using huge bucket dredges which deposit the tailings (the sand and gravel from which the gold dust and nuggets have been washed out) in fantastic patterns.

In order to get the valuable ores, many mining operations must dig out great quantities of worthless rock. Hence, the "dumps" near the head of a mine shaft or below the entrance to a tunnel driven horizontally into a mountain may be much larger than any torrential cones or alluvial fans in the area. Moreover, great piles of rock debris are left behind when ores are put through concentration plants and smelters. On the Witwatersrand in the Transvaal, some sixty million tons of gold-bearing, quartz-rich rock are broken each year, ground to a pulp, and piled in huge hills after the gold is extracted. Some of those hills now stand high above the eucalyptus groves of an otherwise unbroken plain.

With the passage of time, such hills may be covered with vegetation, especially in humid regions, and their true nature be concealed. In West Lothian, Scotland, where oil shale has been mined since the mid-nineteenth

304

century, the spent shale has become grass-covered piles along the railroad and highways that are easily mistaken for natural hills. Even the geologists of the future may have difficulty in identifying all of the man-made landforms of the twentieth century, so vast is the scale and so varied are the results of man's activities as an agent of geologic change.

With the approaching exhaustion of mineral resources close to the surface, man is penetrating deep into the earth's crust to secure raw materials essential to industry and commerce. Mine shafts have been sunk to more than a mile and a half; oil and gas wells have been drilled to depths of approximately five miles. Since the margins of every continent are submerged beneath adjacent seas, the search for mineral wealth now extends far beyond the shores. In northern Nova Scotia, for example, coal mines reach out two miles or more beneath the Atlantic. The same geologic formations and structures responsible for the rich petroleum resources of southeastern Texas and southern Louisiana continue southward for scores of miles beneath the Gulf of Mexico. The water of the Gulf has the same relation to the rocks containing oil as has the veneer of glacial drift to certain oil fields in Ohio or the mantle of sand to the bedrock in large parts of the Libyan and Arabian deserts. An appreciable fraction of United States petroleum now comes from offshore wells in the Gulf of Mexico. Today drilling for oil is proceeding in the North Sea and at other places where the continental shelf is an extension of land areas beneath which the geologic conditions favor the occurrence of oil and gas.

Unexpected things sometimes happen when exploration for mineral resources takes men into little known regions or such places as the floor of the sea. One such episode, for example, occurred during a drilling operation on the continental shelf in Cook Inlet, Alaska, when the drill encountered

Hondsbossche Zeewening dike and its polder, Netherlands. The dike is protected by groins, and the dune in the foreground is anchored by vegetation planted in rows at right angles to the sandy beach. (Netherlands Information Service)

305

A man-made harbor at Los Angeles, California. (William A. Garnett)

gas at a depth of about twelve thousand feet. Due either to the failure of cement to set around the safety valve or the inability of light equipment to cope with the pressure at so great a depth, the gas began erupting through the borehole at a rate of about five million cubic feet per day. The drilling equipment was moved to one side to drill a "relief well" and reduce the pressure so that the original well might seal itself and prevent further loss of gas and condensate oil. In the meantime, the escaping gas was set on fire to prevent contamination of the water and consequent injury to migrating salmon. For days the resulting torch illuminated the waters of the inlet and gave rise to a series of ripples expanding outward from the site of the ruined well.

THE RESOURCES OF THE SEA

Oil and gas from offshore wells, such as the one in Cook Inlet or the many wells in the Gulf of Mexico, are not resources of the sea but of continents that happen at present to be submerged. Resources of the sea are the mineral matter dissolved in seawater or precipitated from it on the ocean floor, and the animals and plants living in the marine environment. There may be valuable mineral resources in the rocks of the earth's crust beyond the continental shelves, but if so, they cannot conceivably be exploited within the foreseeable future.

The mineral matter dissolved in the oceans would make a layer of dried salts about a hundred and fifty feet thick covering the entire globe. Much of

306

this is common salt, NaCl, but there are considerable amounts of magnesium, potassium and bromine, as well as somewhat less of manganese and iron, and still smaller amounts of several other elements, including silver and gold.

Recovering common salt from the ocean by evaporating seawater is man's oldest chemical industry and it is still going on today. Along the coast of the eastern Mediterranean, the Arabian Gulf, and elsewhere in tropical or near-tropical latitudes where the air is dry and the sun hot, are many salt trays or salt pans just above the reach of high tides. Seawater is pumped or carried into them to be evaporated by solar heat. Much of the salt used today for domestic or industrial purposes comes, however, from beds of rock salt among the sedimentary evaporites or from salt domes on the various continents (Plate 114). From these deposits the rock salt is either mined or extracted by forcing superheated water down wells and then pumping it back to the surface as a saturated brine. It will be a long time before the salt resources of the continents are so depleted that it will be more economical to turn to the sea for this essential of food supplies and chemical industries.

In contrast, virtually all of the United States production of magnesium and about four-fifths of its bromine come from seawater. Magnesium is the sixth most abundant element in the earth's crust, but extracting it from such rocks as dolomite or such minerals as magnesite is very expensive. With the development of economical methods for producing magnesium from sea-water, and especially since World War II, the use of this metal in the aircraft, automotive, and many other industries has expanded rapidly. Alloys of magnesium with thorium, zirconium, manganese and aluminum are currently essential for the fabrication of ballistic missiles and satellite launchers. Bromine is chemically very reactive and many important uses have been found for it in recent years. The major part of its current use is as an ingredient (ethyline dibromide) of antiknock gasoline, but much is also consumed in the photographic and pharmaceutical industries. More than a hundred compounds of bromine are used commercially in products ranging from hair-waving preparations and bakery preservatives to fire-extinguishing foams. The life of each of us would be affected if these resources of the sea were not available.

There is no expectation that other metals will soon be extracted from seawater on a commercial basis. They are present in such small amounts per unit of water that the expense of recovery is far greater than their value; they can, moreover, be secured more profitably from land sources. Even bromine is barely within the bounds set by economics; the extraordinarily efficient plant for its extraction at Freeport, Texas, handles about seven hundred million gallons of seawater per year in order to produce fifteen thousand tons of bromine.

The sea may be thought of as a vast laboratory in which many complex chemical reactions constantly take place. In spite of the diluteness of the solutions, some of those reactions result in deposits on the sea floor. Two of these deposits are the manganese and phosphorite nodules which in certain places are sufficiently concentrated to raise the possibility of economic recovery. The manganese nodules also contain minor amounts of nickel, cobalt and zirconium. Oceanographers estimate that the total tonnage is thousands of times greater than the amount available on land. They thus

307

constitute a reserve that may be of crucial importance in the distant future.

At present, the food resources of the sea are far more important than its mineral resources. "Sea food" includes a great diversity of marine organisms—rare delicacies for some people and staple foods for many others. The annual harvest has doubled since 1945 and is now about forty million tons per year. The national diets of the Portuguese, Japanese, Burmese and Indonesians are critically dependent on their fisheries. The sea is a poor source of carbohydrates but an unbelievably rich source of proteins for which a large part of the world's population is presently hungry. Man found it advantageous long ago to cultivate and improve the natural grasses in the best soils he could find; similarly, the cultivation of select oceanic areas is beginning to make equally good economic sense.

It is obviously futile to attempt to transpose agricultural practices from the land to the marine environment, but marine biologists are beginning to understand the mechanisms and processes that concentrate the life of the sea in certain fairly stable regions. Research now under way promises the production of aquatic food on a vastly larger scale. Competent estimates indicate that the sea could meet the protein requirements of many times the present population of the earth.

RESOURCES FOR THE FUTURE

The world population is increasing with unprecedented speed; it doubled between 1900 and 1960 and now totals a little more than three billion. If present trends continue it will reach six billion between A. D. 2000 and 2010. The withdrawal of metallic ores and mineral fuels from the earth's stores is increasing even faster; more have been consumed since 1915 than in all the previous thousands of years of human history. Add the present trend toward greatly increased per capita consumption of mineral resources to the accelerating increase in number of persons and the inevitable question arises: are the resources of the earth adequate to meet the future needs of mankind?

Man's physical needs are three-fold: food for his body; metals and nonmetals for his machines and structures; and mechanical or electrical energy. These needs must be met by using natural resources, either the nonrenewable resources stored in the earth's crust as a result of geologic processes, or the renewable resources due to the force of gravity, such as water power, or to solar radiation, such as the products of the plant and animal kingdoms.

The soils and the bodies of water that support plant and animal life constitute the food factory for mankind. In recent years the efficiency of its operations has been tremendously increased. Wherever men employ techniques and tools of modern science and technology, as in western Europe, Canada, the United States and Australia, food surpluses have become a serious economic and political problem. On the other hand, in many other parts of the world the increase in the food supply has barely kept pace with the increase in population so that there is little or no improvement in living standards and hundreds of millions continue to be undernourished. Education for more efficient land management, agricultural methods, animal

husbandry, and food processing is slow; sufficient capital is not available to obtain the better types of seed, improved breeds of cattle, hogs and poultry, and necessary tools, farm machinery, fertilizers, and pesticides. It is certain, however, that every one of the present inhabitants of the earth could be properly nourished if man's food factory were operated everywhere with the best agricultural procedures, equipment and supplies, and if the foodstuffs thus made available were properly distributed, regardless of economic or political obstacles. The world is one of potential abundance.

How long will it remain so? The increasing number of mouths to be fed can be accommodated to a considerable extent by increasing the number of acres under cultivation. Innumerable swamps and marshes are yet to be drained, extensive arid lands remain to be irrigated, and many more patches of fertile soil can be won from the sea. To enlarge the terrestrial food factory will, however, involve large capital investment and great engineering enterprises. Probably the more important addition to food supplies will come in the immediate future from further increase in productivity per acre on lands already under cultivation and from greater exploitation of the food resources of the sea. None of the sciences and technologies involved in these two possibilities has reached the end of its road.

But there is a limit to the number of people the earth can support. Assume

A strange landscape created near Ester, Alaska, by huge dredges used in the recovery of gold from alluvial deposits. (Steve McCutcheon)

309

the maximum number to be about thirty billion, all of them able to take full advantage of the benefits conferred by the more sophisticated science of the future. Would there be any places left in which man could "commune with nature" if the earth were crowded with ten times its present population? The optimum number is certainly not more than ten billion; it may be as small as five billion, a number that will almost surely be reached during the lifetime of many readers of this book. If something is not done to restrain man's tendency to "be fruitful and multiply," the consequences will be truly calamitous.

Each of the suggested procedures for augmenting man's food supplies entails greatly increased demands upon nonrenewable resources of mineral fuels and of metallic ores. Farms must be mechanized wherever machines will be more efficient than men and animals. Irrigation and flood-control installations, hydroelectric power plants and transmission lines all require power-driven machinery and much iron, steel, copper and other metals for construction and maintenance. (In passing it may be noted that the harnessing of all the potential waterpower of the world would fall short of providing the energy now being used, much less what may be needed in the near future.) Distribution from producers to consumers necessitates a network of transportation facilities—highways and trucks, railroads and locomotives, canals and barges, aircraft and control towers—all of which would increase the demands upon metallic ores and the sources of energy.

There is, however, convincing evidence that the earth's nonrenewable resources are adequate to meet all the demands that will be placed upon them during the next several hundred years. (This gives man plenty of time to learn how to make do with the renewable resources.) The known coal reserves are more than a thousand times the current annual consumption. The known and highly probable petroleum resources are more than two hundred times the maximum annual consumption to date, and even greater quantities of petroleum products can be obtained by distillation of the known deposits of oil shale. Among the mineral fuels must now be included the ores of uranium and thorium, with their fissionable atoms; the amount of energy that may be produced from them is at least five times, possibly ten times, the amount that can be secured from all the coal, petroleum, natural gas, and oil shale in the earth's crust. Thus there is no danger that mankind will exhaust the nonrenewable sources of mechanical and electrical energy in the next thousand years.

For the metallic ores the situation is nearly as reassuring. Rich and easily worked iron-ore bodies were found only a few years ago in Venezuela and eastern Quebec. Newly perfected methods of geophysical exploration are locating other ore bodies completely concealed by overlying rocks. Still more significant, in the Lake Superior region, iron is now extracted economically from the rock known as "taconite," which until the early 1950s was of no value whatsoever, except possibly as crushed stone. If, within the next two or three hundred years, any of the metals seem to be in short supply, new sources will be found, new processes of extraction from lower-grade ores will be invented, or substitutes will be discovered among the other more abundant substances. The earth's hidden mineral wealth should not be discounted.

There is, however, another problem that must not be overlooked. The metallic ores, the nonmetallic deposits of economic value, and the mineral

Above: Drilling for oil near 'Ain Dar in the Arabian Desert. (ARAMCO)

Above, right: A gas "blowout" and drilling barge in Cook Inlet, Alaska. (Steve McCutcheon)

fuels that are basic to modern civilization occur under certain well-defined geologic conditions and in certain places. There is no equivalent in Spain of western Germany's coal and iron, no counterpart of Spain's Almadén mercury ore in Germany. No nation, with the possible exception of the Soviet Union, has within it a sufficient variety of geologic structures to give it adequate supplies of all the metals it needs for modern industrial prosperity. Similarly, no nation enjoys a sufficient variety of climatic conditions to permit all kinds of foodstuffs to be grown on its farms or gathered from its forests or to allow the growth of all the various plants contributing essential raw materials to its industry. Every nation has a long list of "strategic materials" that must be imported from abroad. The world of potential abundance is also a world of inescapable interdependence.

The details of mineral interdependence will change from time to time as new substances find use in new technologies or as replacements for materials already in use. But mineral interdependence will remain a fact of life for centuries to come; indeed it will probably take an increasing hold upon the nations of the world as the deposits of certain minerals are exhausted in one nation or another. The physical barriers that formerly divided mankind into provincial groups are no longer effective; only the economic, social, and political barriers remain. Man's relation to the earth has changed drastically in recent years. To maintain those barriers between regions even so inclusive as an entire continent makes it impossible to take full advantage of the abundance of resources unevenly distributed in and on the earth's crust. Because of its geologic history and structure, the earth on which we live is so fashioned that it may be occupied most satisfactorily by human beings who arrange for the free exchange of raw materials and finished products, of goods and services, and of ideas and ideals the world around. These shall inherit the earth.

GEOLOGICAL TIMETABLE

ERAS	PERIODS	EPOCHS	MILLIONS OF YEARS AGO	PRINCIPAL MOUNTAIN-MAKING EVENTS	GLACIAL EPISODES	EARLIEST RECORDS OF ANIMALS	EARLIEST RECORDS OF PLANTS
CENOZOIC	Quaternary	Recent	0.01				
CENOZOIC	Quaternary	Pleistocene	1.0		The Great Ice Age	Mankind	
CENOZOIC	Tertiary	Pliocene	13	Himalayas Cascades			
CENOZOIC	Tertiary	Miocene	25	Alps			
CENOZOIC	Tertiary	Oligocene	36	Andes			
CENOZOIC	Tertiary	Eocene	58				
CENOZOIC	Tertiary	Paleocene	63				
MESOZOIC	Cretaceous		135	Rockies		Placental Mammals	Grasses and Cereals
MESOZOIC	Jurassic		181	Nevadan		Birds and Mammals	Flowering Plants
MESOZOIC	Triassic		230				
PALEOZOIC	Permian		280	Appalachian, Hercynian and Armorican	Permocarboniferous		Ginkgos
PALEOZOIC	Carboniferous — Pennsylvanian					Insects	Cycads and Conifers Seed Ferns
PALEOZOIC	Carboniferous — Mississippian					Reptiles	Cycads and Conifers Seed Ferns
PALEOZOIC	Devonian		345	Acadian		Amphibians	Ferns
PALEOZOIC	Silurian		405	Caledonian			
PALEOZOIC	Ordovician			Taconic		Fish	
PALEOZOIC	Cambrian		500				
PROTEROZOIC / PRECAMBRIAN			620	Killarney	Early Cambrian	Invertebrates	Mosses Marine Algae
PROTEROZOIC / PRECAMBRIAN					Huronian		
ARCHEOZOIC / PRECAMBRIAN				Algoman			
ARCHEOZOIC / PRECAMBRIAN			1420				
ARCHEOZOIC / PRECAMBRIAN			2300	Laurentian			
ARCHEOZOIC / PRECAMBRIAN			3500				

TIMETABLE OF THE GREAT ICE AGE

PERIOD	EPOCH	NORTH AMERICA Glacial and (Interglacial) Stages	NORTH AMERICA Glacial Substages	YEARS AGO	THE ALPS Glacial and (Interglacial) Stages	THE ALPS Glacial Substages	NORTHERN EUROPE Glacial and (Interglacial) Stages	NORTHERN EUROPE Glacial Substages
QUATERNARY	Recent							
QUATERNARY	PLEISTOCENE	Wisconsin	Cochrane	6,500	Würm		Weichsel	Fennoscandian
QUATERNARY	PLEISTOCENE	Wisconsin	Valders	11,000	Würm	Final Würm	Weichsel	Younger Dryas
QUATERNARY	PLEISTOCENE	Wisconsin	Mankato-Port Huron	13,000	Würm	Late Würm	Weichsel	Older Dryas
QUATERNARY	PLEISTOCENE	Wisconsin	Main Wisconsin	18,000	Würm	Main Würm	Weichsel	Brandenburg-Stettin
QUATERNARY	PLEISTOCENE	Wisconsin	Iowan	40–60,000	Würm	Early Würm	Weichsel	Warthe
QUATERNARY	PLEISTOCENE	(Sangamon)			(Riss-Würm)		(Eem)	
QUATERNARY	PLEISTOCENE	Illinoian		about 200–300,000	Riss		Saale	
QUATERNARY	PLEISTOCENE	(Yarmouth)			(Mindel-Riss)		(Holstein)	
QUATERNARY	PLEISTOCENE	Kansan		about 600–700,000	Mindel		Elster	
QUATERNARY	PLEISTOCENE	(Aftonian)			(Günz-Mindel)		(Tiglian)	
QUATERNARY	PLEISTOCENE	Nebraskan		about 800–900,000	Günz		Älteste	
QUATERNARY	PLEISTOCENE			about 1,000,000				

Glossary

aa. (Pronounced *ah-ah*) Hawaiian term for a rough, clinkery lava. *See* pahoehoe.

ablation. The combined processes that decrease the size of a glacier.

antecedent stream. River that has maintained its course despite the rise of a mountain range athwart it.

anticline. Fold or arch of layers of rock that dip outward in opposite directions from the axis; opposite of syncline.

aphanitic. Pertaining to igneous rocks in which the individual crystals are too small to be seen with the unaided eye.

aquifer. Water-bearing layer of rock, sand, or gravel capable of supplying water to wells or springs.

arête. Sharp crest of a mountain ridge between two cirques or two glaciated valleys.

argillaceous. Pertaining to substances composed entirely or partly of clay.

atoll. Reef or island, encircling a lagoon and built by corals and associated organisms.

backshore. The part of a shore reached by waves during exceptional storms.

barchan. Crescent-shaped dune, with convex side facing the wind and steeper concave slope on the leeward side.

base level. The lowest level to which a land surface can be eroded by running water.

batholith. A huge mass of crystalline igneous rock originating within the earth's crust and extending to great depths.

bergschrund. A deep crevasse near the head of a valley glacier.

bomb, volcanic. A mass of lava ejected from a volcanic vent in a plastic condition and then shaped in flight or as it hits the ground.

caldera. Crater-like depression generally resulting from the subsidence of the central part of a volcano.

cinder, volcanic. A fragment of lava, generally less than an inch in diameter, ejected from a volcanic vent.

cirque. Steep-walled basin high on a mountain, produced by glacial erosion and generally forming the head of a valley.

concretion. A concentration, usually spherical, of mineral matter in sedimentary rocks, produced by deposits from solution.

crust. The outer shell of the solid earth surrounding the mantle.

decke (plural *decken*). *See* nappe.

deflation. The removal of material from a land surface by wind action.

dike. Wall of intrusive igneous rock cutting across the structure of other rocks.

discontinuity. An abrupt change in the physical properties of adjacent materials in the earth's interior.

doline. Funnel-shaped sinkhole resulting from water action on a limestone surface.

drumlin. Oval-shaped hill composed of glacial drift, with its long axis parallel to the direction of movement of a former ice sheet.

eolian. Pertaining to the erosion and the deposits resulting from wind action and to sedimentary rocks composed of wind-transported material.

epicenter. Point on the earth's surface directly above the focus of an earthquake.

erratic. A boulder transported by a glacier or ice sheet.

esker. A long, often winding ridge of glacial drift deposited by a meltwater stream in a tunnel or channel in stagnant ice.

exfoliation. The breaking loose of scales or sheets from rock surfaces by physical and/or chemical weathering processes.

felsenmeer. (German, "sea of rock") Rock waste covering a nearly level area, usually above timberline.

firn. *See* névé.

313

fluviatile. Pertaining to rivers and the sediments deposited in stream beds.

focus. The true center at which the energy of an earthquake is first released.

folia. Close, often wavy bands or laminations, up to four inches in thickness, of unlike mineral composition. The rocks in which they appear are said to be foliated.

foreshore. The part of a shore lying between high- and low-water marks.

fumarole. A vent, usually in a volcanic region, that emits steam or gaseous fumes.

geode. A hollow, spherical body from an inch to a foot in diameter, with a lining of inward-projecting crystals; found mostly in limestones.

geosyncline. A large depression, usually longer than wide, in which sedimentary and sometimes volcanic rocks accumulate to a great thickness while its surface is slowly subsiding.

geyser. An intermittently erupting hot spring.

glacial drift. Boulders, till, gravel, sand or clay transported by a glacier or its meltwater.

glaciofluvial. Pertaining to streams flowing from glaciers and their deposits.

glaciolacustrine. Pertaining to a glacier's deposits in lakes.

glaciomarine. Pertaining to a glacier's deposits in the sea.

gouffre. A large hole opening downward into a cavern in a limestone region.

graded stream. A stream having a smooth gradient, without cascades or rapids.

granitization. An assumed process whereby granite derives from sedimentary rocks without liquefaction into magma.

hoodoo. A grotesque form produced by erosion of rock.

hook. The outer end of a spit that is turned landward by action of waves and currents.

inclusion. Fragment of rock from the wall or roof of a magma chamber enclosed in the rock solidified from the magma.

isoseismal line. A line on a map connecting all points at which a given earthquake had a specified intensity.

isostasy. The principle that, above some assumed level within the earth, different segments of the crust that have the same area also have the same mass and therefore press downward with the same weight; consequently the less the average density of a segment, the thicker it must be and the higher it will stand.

isostatic adjustment. A vertical movement within the earth in response to loading of sediments or of an ice sheet on the crust or to unloading by erosion or melting.

juvenile water. Water from the interior of the earth coming to the surface for the first time.

kame. Conical hill of glacial drift deposited by a meltwater stream against the edge of an ice sheet.

karst. Limestone region with many sinkholes, abrupt ridges, caverns and underground streams. Named for Karst district in Yugoslavia.

laccolith. Lens-shaped body of intrusive igneous rock that has domed up the overlying rocks.

lacustrine. Pertaining to a lake, sediments on a lake bottom, or sedimentary rocks composed of such material.

lahar. Torrential flow of water-saturated pyroclastic debris down the slopes of a volcano.

lapilli. Pellets of lava ejected from a volcanic vent in a liquid or plastic form and shaped in flight.

loess. Sedimentary material, chiefly silt-size particles, but with some clay and fine sand, deposited primarily by wind.

magma. Molten rock material in a liquid or pasty state, originating within the earth at a high temperature.

mantle. The part of the earth's interior beneath the crust and surrounding the central core.

meteoric water. Water occurring in or derived from the atmosphere.

moho. Short name for the Mohorovičić discontinuity believed to separate the earth's crust from the mantle; first identified by a Yugoslav seismologist of that name.

mohole. Drill hole intended to penetrate through the moho into the earth's mantle.

monadnock. Residual hill rising above a peneplain and not yet worn down to that plane; named after Mt. Monadnock, New Hampshire.

moraine. Accumulation of glacial drift with a distinct topographic form such as a ridge.

moulin. Circular depression on a glacier into which meltwater plunges.

nappe. A large body of rock displaced several miles by overthrusting or recumbent folding. Same as *decke.*

névé. Compacted granular snow partly converted into ice. Same as *firn.*

nip. A low wave-cut cliff along a shore, representing the initial stage of wave erosion.

nodule. A small concretion commonly composed of silica.

nuée ardente. Avalanche of fiery ash enveloped in compressed gas from a volcanic eruption.

nunatak. Mountain-top projecting above an ice sheet, as in Greenland.

outcrop. An exposure of bedrock at the surface of the ground.

pahoehoe. Hawaiian term for smooth and billowy or ropy lava. *See* aa.

peneplain. Extensive land surface eroded to a nearly flat plain.

permafrost. Permanently frozen subsoil.

phenocryst. A relatively large and ordinarily conspicuous crystal in a porphyry *(q. v.).*

porphyry. Igneous rock containing conspicuous phenocrysts in a fine-grained or aphanitic groundmass.

pyroclastic. ("fire fragmented") Pertaining to material explosively ejected from a volcanic vent.

rock glacier. Lobate glacier-like mass of rock waste occurring at high altitudes.

rock waste. Fragments of bedrock produced by weathering.

scoria. Slaglike fragment of lava explosively ejected from a volcanic vent.

scree. *See* talus.

seif dune. A long, sharp-crested dune extending in the direction of the wind that constructed it.

seism. An earthquake.

seismograph. Instrument for recording earthquakes.

seismology. The study of earthquakes.

shield volcano. Domelike volcano built chiefly of basaltic lava flows, such as Mauna Loa on Hawaii.

sill. Tablet-shaped body of intrusive igneous rock emplaced parallel to the bedding of the rocks it intrudes.

sliderock. Rock waste that has slid down a slope.

solfataric. Pertaining to sulfurous exhalations from vents in a waning stage of volcanic activity; as in the Solfatara area near Naples, Italy.

solifluction. The movement of soil particles and rock waste chiefly by frost action.

spit. A small point of land projecting into a lake or sea.

stack. Offshore pinnacle of rock left standing when adjacent cliffs have been worn back by wave erosion.

stoping. A process whereby magma works its way up within the earth's crust, engulfing blocks of rock.

syncline. A fold of layers of rock that dip inward from both sides toward the axis; opposite of anticline.

talus. A sloping heap of rock fragments, generally at the foot of a cliff undergoing rapid weathering. Same as *scree.*

tarn. A small rock-rimmed lake in an ice-gouged basin on the floor of a cirque or in a glaciated valley.

tectonic. Pertaining to rock structures and topographic features resulting from deformation of the earth's crust; also earthquakes not caused by volcanic action, landslides, or collapse of caverns.

temblor. An earthquake.

till. Nonstratified glacial drift generally consisting of stiff or sandy clay studded with stones.

tillite. A sedimentary rock composed of firmly consolidated till.

tombolo. A bar connecting one island with another or with a mainland.

tsunami. A sea wave or waves caused by an earthquake. Sometimes incorrectly called a tidal wave.

varve. Two contrasting layers of sediment, commonly clay, one deposited in summer (usually light colored), and the other in winter (usually dark).

ventifact. A stone shaped by the action of wind-blown sand.

Further Reading

GENERAL

Conversation with the Earth. By Hans Cloos. Alfred A. Knopf, Inc., 1953
The Earth. By Carl O. Dunbar. World Publishing Co., 1966
A Planet Called Earth. By George Gamow. The Viking Press, 1963
Understanding the Earth. Edited by I. G. Gass, Peter J. Smith, and
    R. C. L. Wilson. M. I. T. Press, 1971
Earth. By Frank Press and Raymond Siever. W. H. Freeman & Co., 1974
This Sculptured Earth. By John A. Shimer. Columbia University Press, 1959

MORE SPECIALIZED

(The number in parentheses indicates the relevant chapter in this book.)
Plate Tectonics. Edited by John M. Bird and Bryan Isacks.
    American Geophysical Union, 172 (14)
Volcanoes: In History, in Theory, in Eruption. By Fred M. Bullard.
    University of Texas Press, 1962 (12)
A Hole in the Bottom of the Sea. By Willard Bascom.
    Doubleday, 1961 (2)
The Physical Constitution of the Earth. By J. Coulomb and G. Jobert,
    translated by A. E. M. Nairn.   Stechert-Hafner, 1963 (2)
The Changing World of the Ice Age. By R. A. Daly, 1934.
    Reprinted by Stechner-Hafner, 1963 (9)
The World of Ice. By James L. Dyson. Alfred A. Knopf, Inc., 1962 (9)
Glacial and Pleistocene Geology. By Richard Foster Flint.
    John Wiley and Sons, 1957 (9)
Minerals and Man. By Cornelius S. Hurlbut, Jr. Random House,
    1968 (3, 15)
Origin of the Solar System. Edited by Robert Jastrow and A. G. W. Cameron.
    Academic Press, 1963 (2)
Realms of Water. By P. H. Kuenen.
    John Wiley and Sons, 1955 (6, 7, 10)
Causes of Catastrophe. By L. Don Leet.
    McGraw-Hill, 1948 (13)
Marine Geotechnique. Edited by Adrian F. Richards. University of Illinois
    Press, 1968 (8, 14)
Elementary Seismology. By Charles F. Richter.
    W. H. Freeman, 1958 (13)
Volcanoes and Their Activity. By A. Rittman.
    Interscience Press, 1962 (12)
Continental Drift. Edited by S. K. Runcorn.
    Academic Press, 1962 (14)
Landslides and Related Phenomena. By C. F. Stewart Sharpe.
    Columbia University Press, 1938 (11)
Man's Role in Changing the Face of the Earth. By William L. Thomas et al.
    University of Chicago Press, 1956 (15)

Index

Numbers in parentheses indicate color plates; those in italics indicate pages on which black and white illustrations appear.

COLOR ENGRAVINGS BY CHANTICLEER COMPANY. DESIGN AND TYPOGRAPHY BY ULRICH RUCHTI